"An essential call to arms to fight back against the fracking of our minds—and an indispensable guide to reclaiming the human spirit."

—Tristan Harris, co-founder, Center for Humane Technology

"At a time when most reports are of the world getting worse, here's a zinging, erudite book that arrives with the happy news that one thing can get better if we put our minds to it. *Attensity!* is about how to reclaim one of our most powerful and valuable qualities—our attention—through a path back to the human things that matter: community, care, imagination, and art. It's both a keen historical analysis and a call to movement-building from a group of people who have spent years working in the libraries and the classrooms but also, with the shared force of their attention, in the ever-changing streets."

—Nathan Heller, staff writer, *The New Yorker*

"This timely call to action shines a light on the technological and corporate forces that have captured and monetized our attention. Drawing on fields as diverse as neuroscience and philosophy, *Attensity!* offers a brilliant analysis of the fallout of attention capture for selfhood, community, and the environment. An illuminating and liberating read."

—Rob Nixon, author of
Slow Violence and the Environmentalism of the Poor

"Pay attention: If you are human, you must read this book. Also, please note that the term 'attention' has been colonized and made to mean the opposite of what it used to. According to AI people, it now means clearing out context to make less work for pattern-finding algorithms. Don't let algorithms clear YOU out."

—Jaron Lanier, author of *You Are Not a Gadget*

"*Attensity!* is a thrilling declaration of independence from tech's tyranny over our human spirits. We feel the human and humane surge to renewed life through its call to each of us to reclaim ownership of our own attention and of the actions we can cultivate with it—for ourselves and in our own names, instead of at the bidding of machines."

—**Danielle Allen, author of *Our Declaration***

"This is an ambitious text that will demand much of all of us readers beyond the page. It is asking vital questions about the rewiring of our lives in a time of growing crisis, where avalanches of information and access threaten the present and future of care, of close attention. Very thankful to have spent time with this."

—**Hanif Abdurraqib, author of *There's Always This Year***

"*Attensity!* reminds us that how we attend to the world shapes what the world can be for us and for one another. With a lively, even joyful blend of philosophical seriousness and practical imagination, it invites us to see attention not as a private asset to be hoarded but as a shared capacity to be cultivated and protected."

—**Kwame Anthony Appiah, author of *Cosmopolitanism***

"Luminous . . . *Attensity!* is not just a book to be read—it's a call to be answered, a vision to be embraced, a future to be built."

—**James Williams, author of *Stand Out of Our Light***

"*Attensity!* is an unprecedented, impassioned intervention in one of the urgent crises of the twenty-first century. This uncompromising and collectively conceived project is a declaration of tenacious opposition to the ongoing techno-colonization and desolation of our lived time, our attentiveness, and our capacities to care for others, and it fearlessly proposes the creative forms of communal activism and resistance needed to reclaim a shared and livable world."

—**Jonathan Crary, author of *Scorched Earth***

Attensity!
Attensity!
Attensity!
Attensity!
Attensity!
Attensity!
Attensity!

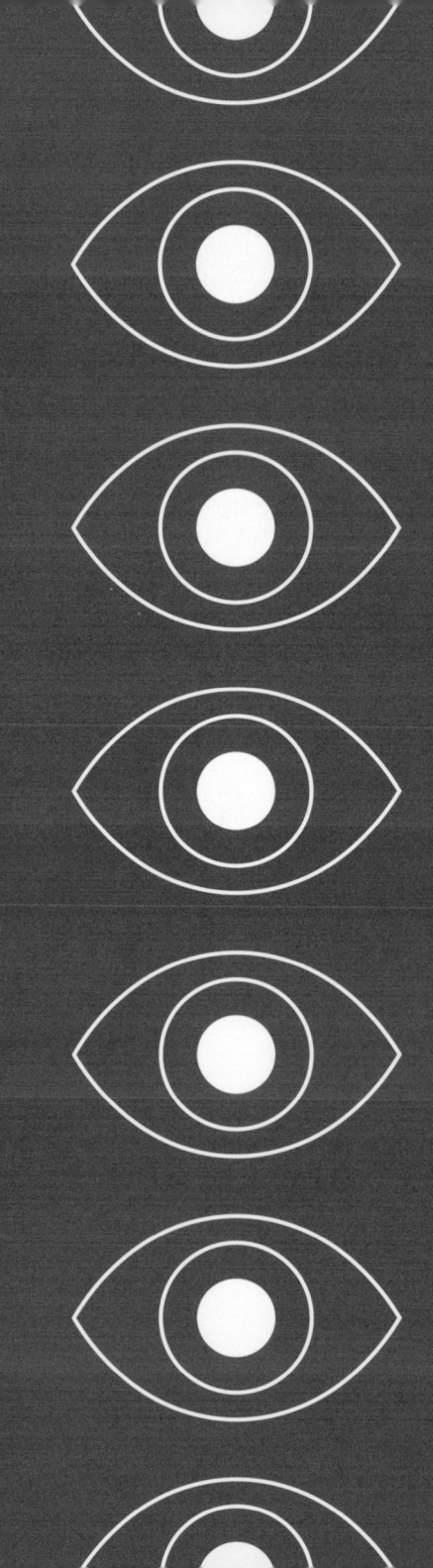

Attensity!

A Manifesto of the Attention Liberation Movement

The Friends of Attention

D. Graham Burnett, Alyssa Loh, and Peter Schmidt, eds.

PARTICULAR BOOKS

UK | USA | Canada | Ireland | Australia
India | New Zealand | South Africa

Particular Books is part of the Penguin Random House group of companies
whose addresses can be found at global.penguinrandomhouse.com.

Penguin Random House UK
One Embassy Gardens, 8 Viaduct Gardens, London SW11 7BW

penguin.co.uk

Penguin
Random House
UK

First published in the United States of America by
Crown Publishing Group, an imprint of Penguin Random House LLC 2026
First published in Great Britain by Particular Books 2026
001

Printed and bound in Italy by LEGO SpA

The authorized representative in the EEA is Penguin Random House Ireland,
Morrison Chambers, 32 Nassau Street, Dublin D02 YH68

A CIP catalogue record for this book is available from the British Library

ISBN: 978-0-241-81096-5

Penguin Random House is committed to a sustainable future
for our business, our readers and our planet. This book is made from
Forest Stewardship Council® certified paper.

MIX
Paper | Supporting
responsible forestry
FSC
www.fsc.org   FSC® C018179

The astonishing reality of things, beings, and persons—
this is the object of pure attention.

—*Twelve Theses on Attention* (Thesis I)

A Few Words About This Book

(and the people who made it)

O ver the last decade a loose network of friends, collaborators, colleagues, and comrades has been thinking and working and making together. The subject? ATTENTION. Human attention, and *inhuman* attention, too. Machine attention (the kind the machines pay to us, and vice versa). Attention as a medium (in art), as a problem (in our shifting technological landscape), as an opportunity (in our lives and in our relationships). Attention as, in deep ways, *the key question of our moment*—in our politics and in our experience of being.

The work in your hands is the work of that community. This is a little unusual, since it isn't easy to write collectively—and it takes time and effort to find a shared voice, and, even more so, a shared vision. But that is what we have tried to achieve in these pages.

Nuts and bolts? We started as an underground association of artists, performers, and interventionists. We created a nonprofit to support work we believed in—work that centered on durational forms of sustained attention in groups (practically speaking, we mostly led educational programs in museums). Then, in 2018, a whole bunch of us ended up in Brazil as part of the São Paulo

Biennial, an event that took place in the immediate wake of the election of Jair Bolsonaro, whose rise was seen by many to reflect the shifting politics of Big Data social media, with its bad actors and bilious cliques. We had come from Poland, Turkey, England, Hungary, Germany, France, and elsewhere, and we were galvanized by the anger we felt in Brazil—and by a shared sense that the "politics of attention" (and specifically the wholesale *commodification* of attention) demanded urgent address.

We reconceived ourselves under a new banner, as the Quaker-inspired "Friends of Attention" coalition, and began a regular series of online meetups, summer school workshops, and annual convenings. We worked out the kinks, and began to create (essays, films, graphic material) together. We read. We wrote. We tried to understand. Some folks fell away. Others joined. We lost one of our dearest to an untimely cancer. Matthew Strother, who died at thirty-five, was, in so many ways, our keel and inspiration (how else to describe someone who slung his chemo pump over his shoulder to dance, and who gave so much of the last year of his life to writing with us—to goading us forward in our solidarity).

In his honor, in the summer of 2023, the Friends of Attention founded the Strother School of Radical Attention in Brooklyn, New York—our flagship center for ATTENTION ACTIVISM, and a sluiceway in the widening surge of the Attention Liberation Movement.

How many are we? The Friends number something like 150 people, we figure—give or take. But we have never counted. It's a coalition of the willing. Now thousands pass through the doors of the School of Radical Attention, to do free "Attention Labs" and participate in our organizing programs and street pedagogy. Maybe thirty people contributed to the writing of this book. And everyone agreed that all the proceeds from its sale would benefit our on-going nonprofit projects.

What do we hope? That this book becomes a kind of *Silent Spring* for our era—a document that can galvanize real change in an area of existential urgency.

Too ambitious? Sure. But that is what kept us writing.

Onward, then! In hope . . .

ATTENSITY!

A Manifesto of the Attention Liberation Movement

You are correct: Something is seriously wrong. It has to do with our ATTENTION, our essential ability to give our minds and senses to the world. This precious capacity has been channeled, captured, and commodified by an industry of immense technological and financial power. How? Call it "human fracking."

Human fracking is bad for people, and for politics. It reduces our very beings (and our relationships) to that which can be quantified, bought, and sold. All this is the triumph of a catastrophic *lie* about what it means to be human. But deceit and exploitation are never inevitable. To push back, we need more than isolated, individual efforts; what we need is a movement of *collective resistance.*

This movement of attentional liberation exists and has a name: ATTENTION ACTIVISM.

Attention Activism is a fight for justice. This emancipatory uprising takes our apocalyptic present, turns it on its

head, and creates, from the chaos and confusion, new vistas of human flourishing.

Attention Activism is rooted in STUDY—a commitment to diverse forms of teaching and learning centered on attention (what it is, what it can be, what it can do). Attention Activism also requires COALITION-BUILDING— collaboration and solidarity across a range of communities who see attention's essential role in human flourishing. Finally, Attention Activism means the formation of SANCTUARIES—spaces where people can gather, care for each other, experiment with different kinds of attention, and conceive brighter futures.

To discern the revolutionary possibilities of the present, we look to artists, thinkers, and dreamers. To bring those possibilities to *bloom,* we heed the countless Attention Activists who are already out there, devising new (and revising old) ways of giving their minds and senses to each other and the world.

These *attentionauts* and *attentionistas* draw on the wisdom of diverse traditions. Across uncharted terrain, emerging practices of joint attention illuminate new horizons of shared political power. Not only power, but beauty, and grace, too.

This is our movement: the free movement of attention in its fullness, freely shared. We call that transformative goodness *ATTENSITY.* Join us in this heightened and heightening glory—or let us join you!

Attensity!
Attensity!
Attensity!
Attensity!
Attensity!
Attensity!
Attensity!

ATTENSITY!

A Manifesto of the Attention Liberation Movement

You are correct: Something is seriously wrong. It has to do with our ATTENTION, our essential ability to give our minds and senses to the world. This precious capacity has been channeled, captured, and commodified by an industry of immense technological and financial power. How? Call it "human fracking."

Human fracking is bad for people, and for politics. It reduces our very beings (and our relationships) to that which can be quantified, bought, and sold. All this is the triumph of a catastrophic *lie* about what it means to be human. But deceit and exploitation are never inevitable. To push back, we need more than isolated, individual efforts; what we need is a movement of *collective resistance.*

This movement of attentional liberation exists and has a name: ATTENTION ACTIVISM.

Attention Activism is a fight for justice. This emancipatory uprising takes our apocalyptic present, turns it on its

head, and creates, from the chaos and confusion, new vistas of human flourishing.

Attention Activism is rooted in STUDY—a commitment to diverse forms of teaching and learning centered on attention (what it is, what it can be, what it can do). Attention Activism also requires COALITION-BULIDING— collaboration and solidarity across a range of communities who see attention's essential role in human flourishing. Finally, Attention Activism means the formation of SANCTUARIES—spaces where people can gather, care for each other, experiment with different kinds of attention, and conceive brighter futures.

To discern the revolutionary possibilities of the present, we look to artists, thinkers, and dreamers. To bring those possibilities to *bloom*, we heed the countless Attention Activists who are already out there, devising new (and revising old) ways of giving their minds and senses to each other and the world.

These *attentionauts* and *attentionistas* draw on the wisdom of diverse traditions. Across uncharted terrain, emerging practices of joint attention illuminate new horizons of shared political power. Not only power, but beauty, and grace, too.

This is our movement: the free movement of attention in its fullness, freely shared. We call that transformative goodness *ATTENSITY*. Join us in this heightened and heightening glory—or let us join you!

You are correct: Something is seriously wrong.

We are living in
conditions contrary
to basic wellness.

W e feel it, too. We, too, are lonely, plagued by anomie, troubled by the sense of a fundamental disconnection between ourselves and *what's real.* The world seems to have slanted sideways. The more we try to be in contact with the people and things that promise to sustain us, the more isolated we become.

But how could it be otherwise? It makes perfect sense that we do not know how to live when our actual people (our friends, co-workers, lovers, even family) increasingly come to us as pop-up snapshots served between advertisements; when our access to the world is determined by our consumer preference categories; when extractive digital networks shape every aspect of daily life; when we do CAPTCHA tests to prove our humanity. We do not know how to live when we are continuously needled by our feeds into anguish and outrage in relation to distant events and lives over which we have no control.

And yet we're *tired* of watching our friends and family (and ourselves) fail to flourish under these conditions. We tire of seeing our people diagnosed with mental disorders—as though failing to flourish in inhuman conditions is some individual quirk, mal-adjustment, or malady.

It is all so obvious that it scarcely bears elaboration: We are living under conditions that are contrary to basic wellness—of our-selves, our communities, and our planet.

We, too, awaken our screens in the morning, and are accosted by headlines announcing epidemics of depression and alienation, the collapse of our most trusted institutions, and the breakdown of our planet's basic ecological cycles. These harms are numerous and diverse, but they are characterized by a common trend toward *dissociation,* toward severing and solitude.

Consider: Over the past ten years, suicide attempts and feelings of persistent hopelessness among high school students have increased by a staggering *40 percent*. Girls are at greater risk—one in ten American female teens actually attempted self-destruction in 2024, and studies suggest that nearly a *third* experienced suicidal ideation. Never before have young people found themselves to be so disconnected from the things that make life worthwhile. We are confronting a pandemic of loneliness, and an unprecedented spike in "deaths of despair."

We see comparable changes in a political landscape ravaged by social rupture and polarization. Instead of the messy-but-hopeful "enactment of democracy," we find ourselves discharging open sewers of contempt into the ever-widening gulf between one virtual silo and another—disintegrating our ability to recognize our common condition, achieve consensus amid conflict, and thereby overcome existential threats.

And the threats *are* existential. The steady advance of climate catastrophe has pitched forward into a full-frontal assault. We are, more than ever, undeniably enmeshed with the worlds of water and air and soil. Yet even as the urgency of our ecological condition batters our spirits with each wildfire, hurricane, or deranging balmy December day, we wonder if, perhaps, this is not so abnormal after all. If our bodily senses return us only to a state of mounting dread, can they be trusted?

What we love most is at stake. It's enough to get anyone out of the chair and into the street!

But among the catastrophes of our present moment is that the very means by which these disasters are brought to our awareness—twenty-four-hour news, the *ping* of cascading notifications, the endless scroll—act not as a bridge into the world, but as the actual mechanisms of further isolation and distance. The more urgent our collective problems become, the *less real* they seem to be.

How is this possible?

How have we been separated from each other and the world—and even from ourselves—at the absolute historical apex of global communicative interconnection?

The answer, we believe, is surprisingly simple: The actual stuff of all connection—the true way we are present to being, beings, and everything else—is neither a network nor a device. It isn't outrage, or, for that matter, hope. It isn't blood, sweat, or tears, either. It is *attention*.

And they have seriously fucked with our attention.

ATTENSITY!

A Manifesto of the Attention Liberation Movement

You are correct: Something is seriously wrong. **It has to do with our ATTENTION, our essential ability to give our minds and senses to the world.** This precious capacity has been channeled, captured, and commodified by an industry of immense technological and financial power. How? Call it "human fracking."

Human fracking is bad for people, and for politics. It reduces our very beings (and our relationships) to that which can be quantified, bought, and sold. All this is the triumph of a catastrophic *lie* about what it means to be human. But deceit and exploitation are never inevitable. To push back, we need more than isolated, individual efforts; what we need is a movement of *collective resistance.*

This movement of attentional liberation exists and has a name: ATTENTION ACTIVISM.

Attention Activism is a fight for justice. This emancipatory uprising takes our apocalyptic present, turns it on its

head, and creates, from the chaos and confusion, new vistas of human flourishing.

Attention Activism is rooted in STUDY—a commitment to diverse forms of teaching and learning centered on attention (what it is, what it can be, what it can do). Attention Activism also requires COALITION-BUILDING— collaboration and solidarity across a range of communities who see attention's essential role in human flourishing. Finally, Attention Activism means the formation of SANCTUARIES—spaces where people can gather, care for each other, experiment with different kinds of attention, and conceive brighter futures.

To discern the revolutionary possibilities of the present, we look to artists, thinkers, and dreamers. To bring those possibilities to *bloom,* we heed the countless Attention Activists who are already out there, devising new (and revising old) ways of giving their minds and senses to each other and the world.

These *attentionauts* and *attentionistas* draw on the wisdom of diverse traditions. Across uncharted terrain, emerging practices of joint attention illuminate new horizons of shared political power. Not only power, but beauty, and grace, too.

This is our movement: the free movement of attention in its fullness, freely shared. We call that transformative goodness *ATTENSITY.* Join us in this heightened and heightening glory—or let us join you!

It has to do with
our ATTENTION, our
essential ability to
give our minds
and senses
to the world.

In efforts to define attention, some emphasize its relationship to catalyzing action, while others emphasize its endless receptivity. We resolve the tension between these two definitions by offering a third.

W hat is attention? The experts do not agree. In fact, they actively contradict each other.

These contradictions affirm the urgency of our cause. They are evidence that attention is already *a site of active contestation*. To affirm what attention is, or is not, or may be, is to decide what sort of world(s) we want to live in.

So let us consider two sides of the matter.

On the one hand, we find a definition of attention that has been *literally instrumental* in service of (and which has developed contemporaneously with) the attention-sucking regime we seek to resist. Now a classic on business school syllabi across the country, management consultant gurus Davenport and Beck's blowout bestseller, *The Attention Economy* (2001), says this: "Attention is what catalyzes awareness into action."

Plenty to admire. After all, awareness and action are central to our practice of attention (and to our movement!). In "catalyze," we may even detect a whiff of attention's obscure alchemical powers.

But on the whole, this formulation is rigid and mechanistic. If awareness is not attention until we act upon it, then our attentional life is therefore only ever a *cascade of incitements*. Davenport and Beck's account sets aside the question of "what attention can be" and replaces it with an always-already-*instrumentalized* definition: Attention is a task-directed TOOL. To be used. It is reflex and trigger. Track and click.

In perfect contrast, let us consider Bernard Stiegler's *Taking Care of Youth and the Generations* (2008). Stiegler, a French philosopher whose work centers on technology, approaches the problem etymologically, writing that "to pay attention is essentially to wait (*attendre*)."

Wait for what? His answer to this goes to the heart of his

philosophy, so it is difficult to do it justice, but the core can nevertheless be briskly stated: "What attention . . . waits on/for . . . is the infinity of the object"—an infinity, he argues, that, mirror-like, *reflects the inner infinity of whoever is attending.*

(Pause, for just a moment, to *feel* what Stiegler is getting at: that within each of us, within YOU, is an infinity that can be perceived through deep attention to the world around us. That to give attention to objects and to other people is to be present to one's ongoing and unending coming-into-being. This coming-into-being happens in a relational space between the "I" and the beings and entities we encounter in our daily lives—*precisely* the space that has been partitioned, and quantified, and optimized, and strapped to a system of psychic extraction that fractures our minds and plunders our eyeballs.)

But that is getting ahead of ourselves. What we can say here with certainty is something like this: According to Stiegler, **attention is a species of (possibly eternal)** *waiting.*

You may detect in these words a familiar set of images: the Zen practitioner atop her zafu; the Catholic taking Communion; the chavruta's weeks-long exegesis of a single Talmudic phrase. This is no mistake—the great spiritual traditions (of which we have named only a few) represent deep and rich attentional worlds realized by diverse practices of waiting upon the infinite. (If you identify with these traditions, then we believe that you *already* share our core commitments. What's more, we think you have much to contribute to Attention Activism.)

So "waiting" is not an altogether bad formulation. We admire Stiegler's deferral of reflex gratification. We admire that his definition is less an incitement than an invitation, for we are in the business of invitation. But at the end of the day, Attention Activists are a coalition committed to action—and the emergent mysteries of indefinite expectancy *do not a revolutionary movement make!*

And so, we come to a contradiction. These ideas seem thoroughly at odds with each other, even formally antithetical. On the one hand, attention as the action-trigger; on the other hand, attention as a kind of perpetual suspension. Which is *right*?

We propose a "resolution" of this paradox. A détente if you will. It requires a slide sideways, out of the realms of business and philosophy and into the world-building openness of literature—or, if you like, life itself.

In his 1902 novel, *The Wings of the Dove,* Henry James depicts a crucial, if fleeting, encounter between a fatally ill patient (female, sensitive, anguished) and an esteemed medical doctor (grand, humane, busy). It is a charged rendezvous, and a rushed one. For various reasons, they will have only a few moments together—the time must be stolen from the exigencies of ordinary life and obligation.

They sit. And here is how James evokes the redemptive power of that moment, making use of the language of attention, and figuring it as a gift of pure imminence: "So crystal clean, the *great empty cup of attention* that he set between them on the table."

This isn't quite indefinite waiting, although there is something of the infinite in the cup's emptiness. And it isn't triggering action either, although we sense in the simplicity of the doctor's gesture the possibility that the cup is already, somehow, unaccountably *full.* There is presence here, a welcoming, an invitation that is also a generous and vital *offering.*

In this fashion, James's account rescues attention from the jaws of antinomy. Attention is no mere tool; it creates a space beyond the stultifying operational logics of technology and capital. Yet it is not an endless, sublime adjournment, either. It moves in the world of frailty and pain. It bears the promise of healing—or at least of consolation. It is the true gift of the open. It is where we meet in that openness, and make space for what unfolds.

This may be the best account we have of what attention can be if it is to be truly ours, if it is to be the stuff of care, and the ethereal medium out of which we make our relationships—to ourselves, to others, and to the world.

The radical philosopher Simone Weil famously wrote, shortly before her death, that "attention, taken to its highest degree, is the same thing as prayer." She said that, in its purest form, true attention presupposes both faith and love.

Faith. Love. We will need these notions as we go about deciding *what it is that we seek to protect.*

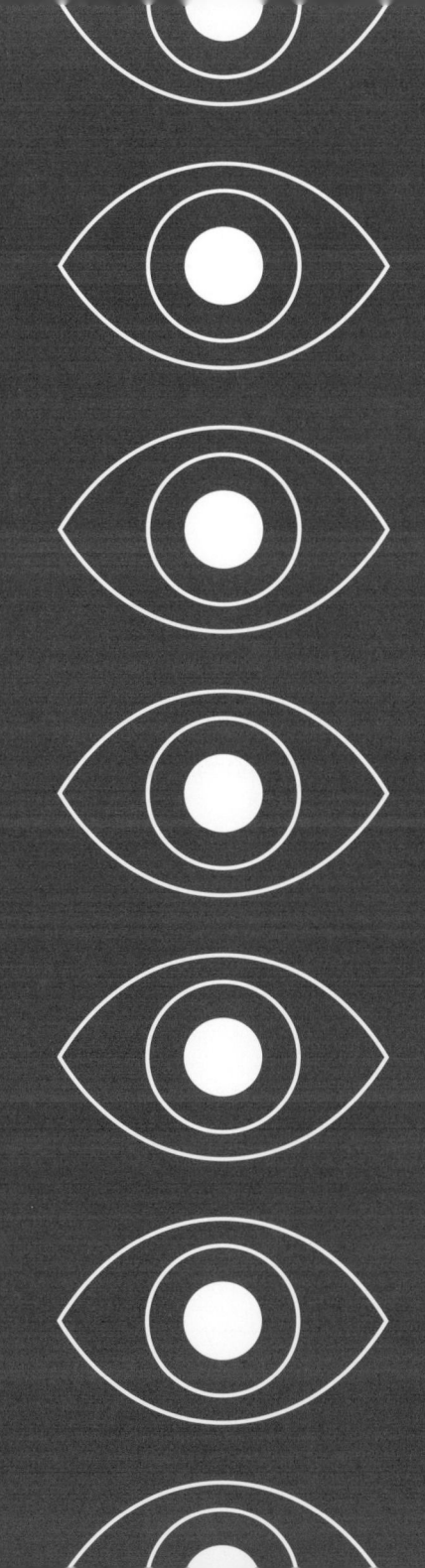

ATTENSITY!

A Manifesto of the Attention Liberation Movement

You are correct: Something is seriously wrong. It has to do with our ATTENTION, our essential ability to give our minds and senses to the world. **This precious capacity has been channeled, captured, and commodified by an industry of immense technological and financial power.** How? Call it "human fracking."

Human fracking is bad for people, and for politics. It reduces our very beings (and our relationships) to that which can be quantified, bought, and sold. All this is the triumph of a catastrophic *lie* about what it means to be human. But deceit and exploitation are never inevitable. To push back, we need more than isolated, individual efforts; what we need is a movement of *collective resistance.*

This movement of attentional liberation exists and has a name: ATTENTION ACTIVISM.

Attention Activism is a fight for justice. This emancipatory uprising takes our apocalyptic present, turns it on its

head, and creates, from the chaos and confusion, new vistas of human flourishing.

Attention Activism is rooted in STUDY—a commitment to diverse forms of teaching and learning centered on attention (what it is, what it can be, what it can do). Attention Activism also requires COALITION-BUILDING—collaboration and solidarity across a range of communities who see attention's essential role in human flourishing. Finally, Attention Activism means the formation of SANCTUARIES—spaces where people can gather, care for each other, experiment with different kinds of attention, and conceive brighter futures.

To discern the revolutionary possibilities of the present, we look to artists, thinkers, and dreamers. To bring those possibilities to *bloom*, we heed the countless Attention Activists who are already out there, devising new (and revising old) ways of giving their minds and senses to each other and the world.

These *attentionauts* and *attentionistas* draw on the wisdom of diverse traditions. Across uncharted terrain, emerging practices of joint attention illuminate new horizons of shared political power. Not only power, but beauty, and grace, too.

This is our movement: the free movement of attention in its fullness, freely shared. We call that transformative goodness *ATTENSITY*. Join us in this heightened and heightening glory—or let us join you!

This precious capacity has been channeled, captured, and commodified by an industry of immense technological and financial power.

The five largest companies in the world, each valued at over two trillion dollars, share a core business operation: the capture and monetization of attention.

T hings did not have to take this shape. It bears saying up front—*things could have been different.* But the fact of the matter is this: Over the last thirty years, the explosive growth of the internet and the global proliferation of data devices have converged with a series of shortsighted political decisions; along the way, the shrewd initiative of a generation of tech entrepreneurs has produced, in effect, an ostensibly "free" digital universe—*whose hidden operating cost is the depletion and pollution of the minds and senses of its users.*

Simply put, most of the internet runs on a business model that is bad for us. And everyone basically knows this. We call it the "attention economy," but in practice it is a globe-spanning industrial farm that extracts money from a billion vegetative humans suspended in an infinite web, eyes glazed. Remember when Neo wakes up in the embryonic slime-pod in *The Matrix*? Like that. Only before he wakes up.

Yes, there are, predictably, a few tech apologists who try to argue that the benefits of this arrangement outweigh its risks. But they mostly don't let their own kids have phones until they go to college, and make them study violin (with tutors, in fortress-mansions). More significantly, the whole idea of treating the situation as a cost-benefit analysis ignores the fact that our *dehumanization cannot be quantified.* This is not a conversation about trade-offs. This is a conversation about coercion, theft, and the instrumentalization of human life.

We spend nearly the entirety of our waking hours in digital spaces substantially financed by extractive profit models that systematically tap human beings for the money-value of our eyeballs. This twelve-trillion-dollar operation drives a global-scale, computationally intensive, and commercially lucrative system for the sourcing, aggregation, and nonstop auction-market of human attention—an industry of psychic extraction that has become, in the twenty-first century, what slavers and pirates were to an earlier age: *hostis humani generis,* Latin for "an enemy of humankind"; which is to say, not an enemy of this person, or that person, or even of this nation or that, but an enemy of *all of us,* in our actual humanity.

At the time of our writing, the five largest companies in the world are each valued at more than two trillion U.S. dollars. All five of these are technology companies (Apple, Nvidia, Microsoft, Alphabet [i.e., Google], and Amazon), and four of the five have *billions* of customers who use their products more or less continuously, and whose data-commodification constitutes a central business operation. The fifth makes the chips that power the processing and manipulation of that data, increasingly by means of algorithmic AI. The point of all of this, of course, is *making money.* This Big Tech nexus makes use of its inescapable digital platforms to collect information about us—our clicks and keystrokes, our blinks and grimaces, our reactions and distractions, our typos and our personality types—and then feeds these data into immensely powerful predictive technologies. The presumed tastes and tendencies of our future selves are then sold to the highest bidder. All of this happens out of sight. It never stops.

The high-octane commercial battle for "eyeballs" that has played out across our digital ecology over the last fifteen years—call it the attention economy, or see it in the still wider frame, as "surveillance capitalism"—long ago gave rise to a demented behaviorist arms race in which only the fastest, loudest, and most explosive content can strike home and hold a fleeting moment in our consciousness. And holding moments in consciousness is the ultimate online moneymaker. The result? We traverse daily a continuous minefield of mini-monstrosities where our attentional capacities are relentlessly broken down into ever more fleeting fragments.

It is indeed a *race to the bottom of the brain stem,* as the philosopher James Williams has pointed out, but what seizes our eyes isn't always violence or pornography. Sometimes it's a viral dance video, or a suggestive headline, or a simple *like* from a friend. But it never stops coming, and it comes at us *all the time*: always surprising, always bright, always loud, always grabbing the gaze. We *like* a lot of this stuff. We *choose* it (or feel like we do—at least until we look up, hours later, and feel weirdly ill). But we cannot forget: We are, all the while, pawns in a money-game of literally *inconceivable* scale.

For all the moments of fun, actual damage creeps in along the whole way. No matter our best intentions, we are trapped. We cannot look away. The intimate and turbocharged intelligence of the machines has wormed its way into our innermost drives and needs, interposing itself between our conscious minds and ... *EVERY-THING WE CARE ABOUT.* And from that extraordinary, unprecedented position these systems exploit us. For cash.

In a real sense, *our attention does not belong to us.* It is, every single day, being run by the ghosts in the machines. They were built for this.

ATTENSITY!

A Manifesto of the Attention Liberation Movement

You are correct: Something is seriously wrong. It has to do with our ATTENTION, our essential ability to give our minds and senses to the world. This precious capacity has been channeled, captured, and commodified by an industry of immense technological and financial power. How? **Call it "human fracking."**

Human fracking is bad for people, and for politics. It reduces our very beings (and our relationships) to that which can be quantified, bought, and sold. All this is the triumph of a catastrophic *lie* about what it means to be human. But deceit and exploitation are never inevitable. To push back, we need more than isolated, individual efforts; what we need is a movement of *collective resistance*.

This movement of attentional liberation exists and has a name: ATTENTION ACTIVISM.

Attention Activism is a fight for justice. This emancipatory uprising takes our apocalyptic present, turns it on its

head, and creates, from the chaos and confusion, new vistas of human flourishing.

Attention Activism is rooted in STUDY—a commitment to diverse forms of teaching and learning centered on attention (what it is, what it can be, what it can do). Attention Activism also requires COALITION-BUILDING— collaboration and solidarity across a range of communities who see attention's essential role in human flourishing. Finally, Attention Activism means the formation of SANCTUARIES—spaces where people can gather, care for each other, experiment with different kinds of attention, and conceive brighter futures.

To discern the revolutionary possibilities of the present, we look to artists, thinkers, and dreamers. To bring those possibilities to *bloom*, we heed the countless Attention Activists who are already out there, devising new (and revising old) ways of giving their minds and senses to each other and the world.

These *attentionauts* and *attentionistas* draw on the wisdom of diverse traditions. Across uncharted terrain, emerging practices of joint attention illuminate new horizons of shared political power. Not only power, but beauty, and grace, too.

This is our movement: the free movement of attention in its fullness, freely shared. We call that transformative goodness *ATTENSITY*. Join us in this heightened and heightening glory—or let us join you!

Call it "human fracking."

These companies
heedlessly pollute
our inner environments
in order to capture
and sell as much of our
attention as possible.

The harms of this new economy cannot be overstated. Human attention is the stuff out of which we care for ourselves, our communities, and our planet. When it is fractured and polluted, all forms of life suffer.

And "fracking" is the right word for the underlying dynamic of the attention economy. In the world of fossil fuel extraction, it refers to the (relatively new) technique by which difficult-to-access oil and gas can be brought to the surface and turned into money. How? Instead of just tapping a nice rich gusher of crude sitting in an easy-access, high-pressure geological boil right under the surface, the frackers must suck their now *very* hard-to-reach bounty out of acres of spongy rock deep in the earth. They pull this off by pumping millions of gallons of high-volume, high-pressure chemical detergent into the ground, which forces the good stuff up to the surface.

The analogy is perfect: The money-value of our attention is "low grade," in the sense that, in aggregate, you can certainly cash it in, but you need a *lot* of human eyeballs to make any profit in the advertising industry. And salable eyeball attention isn't that easy to get, either. *Homo sapiens* have been going about their business being interested in the world and each other for about two hundred thousand years. Getting them all to look at ten square inches of backlit glass for eight or nine hours a day *isn't that easy.*

So the new industrial-scale practitioners of human fracking have to pump vast quantities of high-volume, high-pressure (toxic) slurry into our faces, at the scale of the whole *population,* to drive a bit of biddable attention to the surface.

One can lean on the analogy even harder. After all, fracking the deep earth to extract valuable hydrocarbons and fracking the deep structure of the human mind to extract marketable attention have both emerged as invasive, destructive technologies of wealth-creation across decades of fierce and largely unregulated competition in cutthroat industries. It is possible to trace parallel trajectories: Forty years ago, anxieties about global petroleum and natural gas resources reached market-distorting proportions. People genuinely worried that there wasn't enough accessible oil in the ground to permit the world (and the black gold industry) to go about its business for the foreseeable future. Enterprising futurists turned their attention to alternative forms of energy and civilizational patterns less dependent on fossil fuels. The perfection of fracking techniques has given the legacy corporate power of Big Oil a new lease on life, and at present, despite mounting anxieties about the dangers of anthropogenic climate change, there's really no end in sight for cheap fossil fuels. Fracking has turned out to be a way to perform serial cesarean sections on the goose that lays golden eggs—it's ugly, and the goose isn't going to live forever, but the eggs do stack up surprisingly fast.

Something similar went down across the last half century in the advertising industry. The heyday of *Mad Men*–style Madison Avenue swagger was indeed back in the 1960s. The rise of the tech and financial sectors in the '70s and '80s relegated the world of advertising to a rather unglamorous side hustle in the global money vortex. Human fracking has changed that tune, catapulting the project of extracting money-value from human eyeballs to the veritable forefront of global plutocratic endeavor.

Is there a lesson in all this? Sure. Human beings, going about their business, striving relentlessly to stash away some bucks, will continuously come up with new ways to get their hands on a little bit more of this and that. Leave us to ourselves and we are ingenious, individually and collectively—clever at seizing, adroit at amassing, and relatively heedless as to the consequences. We will destroy for money—sometimes quickly and savagely, but more often slowly, subtly, surreptitiously, with a smile on our faces and a fond hope that no one will notice—until we've got ours. The medium-term effects of this dynamic are never pretty.

In the case of the entrepreneurial frackers, we are inheriting parallel wreckage: Petroleum fracking is doing irreversible damage to our *external* environment (our woods and fields, our water and sky) while the human frackers are verifiably destroying our *interior* environment (our minds and hearts, our ability to be and to sit with ourselves and the people we love). These inner and outer environments depend upon each other. Without caring for the former, we cannot do the hard political work of caring for the latter. Nothing less than the future of the world (and *our* future as non-inhuman beings) is at stake.

ATTENSITY!

A Manifesto of the Attention Liberation Movement

You are correct: Something is seriously wrong. It has to do with our ATTENTION, our essential ability to give our minds and senses to the world. This precious capacity has been channeled, captured, and commodified by an industry of immense technological and financial power. How? Call it "human fracking."

Human fracking is bad for people, and for politics. It reduces our very beings (and our relationships) to that which can be quantified, bought, and sold. All this is the triumph of a catastrophic *lie* about what it means to be human. But deceit and exploitation are never inevitable. To push back, we need more than isolated, individual efforts; what we need is a movement of *collective resistance.*

This movement of attentional liberation exists and has a name: ATTENTION ACTIVISM.

Attention Activism is a fight for justice. This emancipatory uprising takes our apocalyptic present, turns it on its

head, and creates, from the chaos and confusion, new vistas of human flourishing.

Attention Activism is rooted in STUDY—a commitment to diverse forms of teaching and learning centered on attention (what it is, what it can be, what it can do). Attention Activism also requires COALITION-BUILDING— collaboration and solidarity across a range of communities who see attention's essential role in human flourishing. Finally, Attention Activism means the formation of SANCTUARIES—spaces where people can gather, care for each other, experiment with different kinds of attention, and conceive brighter futures.

To discern the revolutionary possibilities of the present, we look to artists, thinkers, and dreamers. To bring those possibilities to *bloom*, we heed the countless Attention Activists who are already out there, devising new (and revising old) ways of giving their minds and senses to each other and the world.

These *attentionauts* and *attentionistas* draw on the wisdom of diverse traditions. Across uncharted terrain, emerging practices of joint attention illuminate new horizons of shared political power. Not only power, but beauty, and grace, too.

This is our movement: the free movement of attention in its fullness, freely shared. We call that transformative goodness *ATTENSITY*. Join us in this heightened and heightening glory—or let us join you!

Human fracking is bad for people, and for politics.

Psychological distress and political discord are on the rise. By the time that researchers can prove a connection to human fracking, it will be too late. We can trust our senses, trust our judgment—and act now.

Here's a thought experiment: What if we were to think of our attention as an absolutely essential precondition of life itself? Because it is *exactly this*. So, what if we were to think of it as something like, say, WATER? A fundamental resource without which our beings cannot survive. Would not, then, the existence of a vast network of thieving siphons seem like a nightmare out of some *dystopic science fiction*?

We would need to imagine a Lovecraftian remake of *Little Shop of Horrors* in which all the parasitic plants of the world join forces in a world-spanning web of sucking extraction. Oh, wait! We don't need to imagine that. Because that is the bona fide *world in which we live*. We daily confront an *existence-spanning web of sucking extraction*, which taps us not for water, but for the *soul-water* that is attention—for what is soul-water if not our ability to *care*, our actual *caring itself*, by means of time and mind and touch and all the senses and *capacities for thought*, too? And what is attention if not these things?

The biologists have a special word—a lovely-scary word—for the sinister contraptions by which parasitic organisms sap and tap and suck at their hosts. Those little tubules and sharp-tipped thready siphons are known as HAUSTORIA, and they do indeed look like something out of a nightmare. In the polite language of science, they are the tentacular processes by which the predatory organism "mechanically invades the host" through a combination of "enzymes and brute force." The enzymes do the careful work of dissolving the membranes and protective defenses of the prey, and "brute force" drives the suckers into its tissue. Etymology always tells tales: The term "haustoria" hails from the Latin word for DRINK, and it means, in effect, "the DRINKERS." For that is what those little tubes are doing: They are *drinking* from the body juices of their neighbor. We throw the term "bloodsucker" around pretty casually, and everyone thinks they know what a vampire does. But actual vampire bats just use a sharp tooth to make a little poke, and lap up a few drops of the good stuff—not much more than would end up on a Band-Aid. Whereas the haustoria are genuinely like an IV in reverse, sucking all the time—just like the human frackers, who have got their bright little pipelines plugged into the faces of everyone you know. And you, too.

How bad is this, really? After all, people seem to like looking at stuff on their phones. Nobody is holding a gun to their heads—or at least not very often. Mostly we all would sorta weirdly rather be on our devices most of the time (checking who texted us, scrolling through Instagram or TikTok, playing a game). Large corporations also make money by selling breakfast cereal, but this doesn't feel like a threat to humanity.

So is there any *real evidence* that we are being systematically "dehumanized" by the attention economy?

The short answer is "Yes, there is." If you are taking the time to read this book, you might already be aware that the U.S. Surgeon General recently issued a national report warning that social media may present serious risks to youth mental health. The American Psychological Association has likewise released a health advisory on kids and online addiction. If you want a bunch of statistics, go online, and start reading around. The psychologist Jonathan Haidt and his colleagues have assembled a series of open-access essays that compile and discuss *hundreds* of peer-reviewed scientific papers addressing a set of basic harms in quantitative and qualitative terms: the impact of algorithmically driven, ad-financed social media ecosystems on education, political dysfunction, and so on. These synopses run for hundreds of pages, and offer many links to the original papers.

But here is the thing: You can also find experts (academics, people with advanced degrees and university positions) who will argue with you about basically any aspect of all that data—all that research. We talk more, below, about the larger history of scientific controversies, and we situate the current "debate" with respect to a set of earlier moments when the best empirical practices took a *very long time* to achieve consensus on something TOTALLY OBVIOUS (cigarette smoking and cancer, anyone?).

Skip ahead if you are especially hungry for that discussion. Tipping our hand, we'll just say here that proving causality is hard, statistics are a messy business, and science is always "historical"—meaning it is always happening in a particular time and place. This means it is always entangled in complicated situations and contingencies. We are pro-science! But waiting until the scientists have *PROVED that we have been dehumanized* is going to mean that we've WAITED TOO LONG.

So use some common sense. Check yourself. How do you feel? What about your friends? Look around. See anything odd? Like, everyone staring at a device, and nobody talking? Nobody *actually seeing* each other? Or, for that matter, the *world*? Or *anything* but a stream of "content" that is continuously monetizing their consciousness? Hmm.

And sure, there are statistics. One in nine American kids has been clinically diagnosed with an attentional disorder. Lots of the doctors think that is an *undercount*. Perhaps more tellingly, about a quarter of adults believe they, too, have such a condition. Half of American teens report that they are on networked devices "almost constantly." Undisputed facts like these have led Australia to take the bold step of completely banning social media use for those under the age of sixteen.

Meanwhile, by every index, the United States is currently riven by levels of discord and mutual distrust unseen since the calamity of the Civil War. Nearly 85 percent of Americans from across the political spectrum believe political discourse has become less respectful, and nearly 80 percent think it has become less truthful, too. A majority of us say that politics makes us "angry," and if you let people pick one word to characterize the political life of the Union, the term we choose more than any other is "divisive" ("corrupt" runs a close second).

Does anyone really want to argue that these developments are unrelated to the dramatic transformation of the public sphere across the last thirty years? After all, Americans now get their news from social media platforms, which are themselves built on, and financed by, the human fracking project. Fracking the public forum has turned out to be very good business—and *very bad* for the civic conditions of democratic politics.

For decades, reputable scientists held out and weren't *absolutely* sure about human-caused climate change. For decades, there remained an "on the one hand, on the other hand" program of rapportage about global warming. Now, finally, there is consensus. And, well, it may actually be too late.

Let's turn our attention to our attention without waiting.

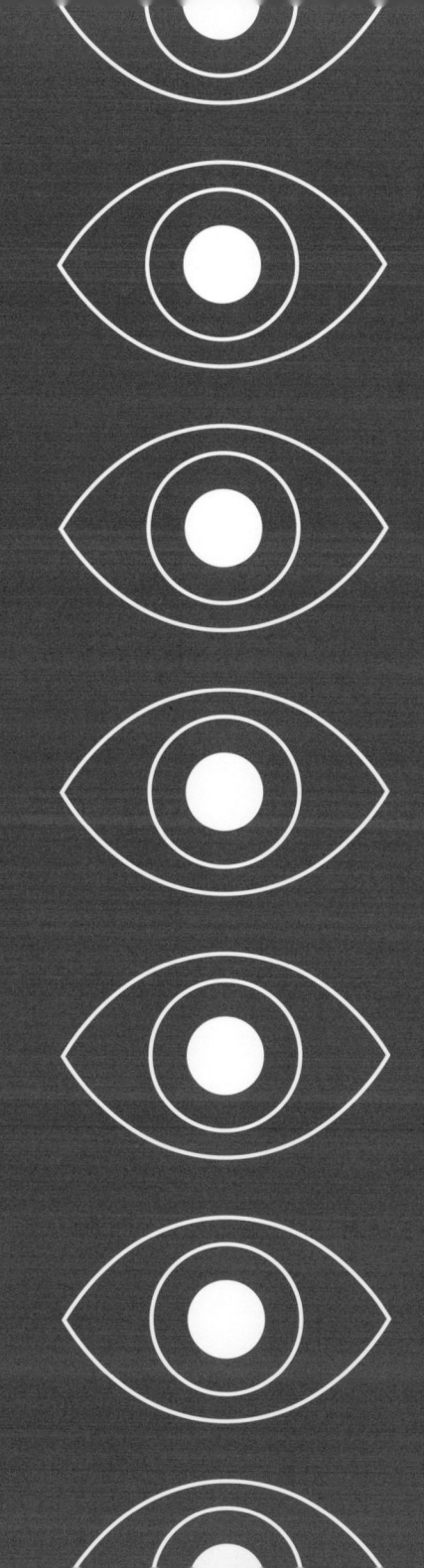

ATTENSITY!

A Manifesto of the Attention Liberation Movement

You are correct: Something is seriously wrong. It has to do with our ATTENTION, our essential ability to give our minds and senses to the world. This precious capacity has been channeled, captured, and commodified by an industry of immense technological and financial power. How? Call it "human fracking."

Human fracking is bad for people, and for politics. **It reduces our very beings (and our relationships) to that which can be quantified, bought, and sold.** All this is the triumph of a catastrophic *lie* about what it means to be human. But deceit and exploitation are never inevitable. To push back, we need more than isolated, individual efforts; what we need is a movement of *collective resistance.*

This movement of attentional liberation exists and has a name: ATTENTION ACTIVISM.

Attention Activism is a fight for justice. This emancipatory uprising takes our apocalyptic present, turns it on its

head, and creates, from the chaos and confusion, new vistas of human flourishing.

Attention Activism is rooted in STUDY—a commitment to diverse forms of teaching and learning centered on attention (what it is, what it can be, what it can do). Attention Activism also requires COALITION-BUILDING—collaboration and solidarity across a range of communities who see attention's essential role in human flourishing. Finally, Attention Activism means the formation of SANCTUARIES—spaces where people can gather, care for each other, experiment with different kinds of attention, and conceive brighter futures.

To discern the revolutionary possibilities of the present, we look to artists, thinkers, and dreamers. To bring those possibilities to *bloom*, we heed the countless Attention Activists who are already out there, devising new (and revising old) ways of giving their minds and senses to each other and the world.

These *attentionauts* and *attentionistas* draw on the wisdom of diverse traditions. Across uncharted terrain, emerging practices of joint attention illuminate new horizons of shared political power. Not only power, but beauty, and grace, too.

This is our movement: the free movement of attention in its fullness, freely shared. We call that transformative goodness *ATTENSITY.* Join us in this heightened and heightening glory—or let us join you!

It reduces our very beings (and our relationships) to that which can be quantified, bought, and sold.

The human frackers are exploiting one very particular kind of attention, one that comes out of a century-old system of laboratory research, which studied attention for specific and narrow purposes (i.e., military performance, advertising, factory productivity). But there is much more to human attention than that.

E very system of exploitation has an internal logic. For all its complexity, the trillion-dollar human fracking industry is the extreme application of a simple idea: that attention can be quantitatively measured (and, by extension, bought and sold). Recognizing this idea and understanding where it comes from can help us to push back against the catastrophic harms that it has produced.

Let's go right at it.

Over the past century or so, a VAST RANGE OF HUMAN FACULTIES AND EXPERIENCES has been effectively *collapsed* into a single, narrow, cybernetic (i.e., machine-engaged), quantifiable, and ultimately *always-already-instrumentalized* THING— a thing we CALL "attention," and worry about as "attention," but that has as little to do with the true *cosmos* of human attention as a sex-ed class has to do with human love. Which is to say, not nothing. But not that much, either.

This narrow way of thinking about "attention," which arose in the experimental psychology laboratories of the long twentieth century, has come to shape the way that all of us think about (and use) our attention. To put it bluntly, human attention was sliced and diced in those laboratories, en route to being *priced* in the marketplace of the attention economy.

To understand this important point requires a turn into the history of science. After all, when we want "facts" about attention now, to whom do we turn? Research scientists. And what do they say "attention" is? Well, they are hardworking people. And they have studied diligently for years, and they know a lot. They have machines that are expensive and complicated, and they can point to a screen and show you a diagram that looks like a lava lamp (it's a real-time image of your brain in action) and say things like "There, that's the attention happening!"

All of this is fantastic. No objections whatsoever. All the stuff they say is attention, and the stuff they say about it—all of it is, we are sure, "right." Indeed, if we could get down into the weeds and argue with them about it, we'd be *OTHER scientists studying attention.* And we are not (or at least *most of us* aren't). So, by definition, what they have to say is, for now, "what you can say about attention" that is backed up by the best new scientific findings.

We aren't going to review what they say here, because, again, there is a LOT of this stuff out there on the internet, and you are welcome to go watch a few videos of reputable scientists saying stuff about what attention is and how it works (tons of interesting things!—"attentional blindness," "feature integration theory," top-down vs. bottom-up processing problems, etc., etc.). Plus, there are many good popular science books about attention—some easy, some hard. Jump in and learn some stuff. We support! Not least because, of course, you'll be training your attention as you *study.* (More on that, below.)

For now, however, we are going to get *perpendicular* to that whole plane. We are going to rise up and view that whole scientific conversation about attention from the lofty vantage of historical distance, and in doing so, we are going to remind ourselves that our attention is *OURS,* and it does not belong to anyone else—not even Nobel laureate neurophysiologists. Here's the key thing: No knowledge of attention from outside, no matter how fancy or technical, can ever substitute for the *immediacy and experiential reality of YOU,* and what it is like for you to be in the world, and give yourself to it, and let it come to you.

In fact, let's start right there. *That* activity, whatever it is, that "what it's like for you to be in the world," was NOT actually what the twentieth-century experimental scientists studying attention got grants to research. Nope. They got grants to study whether a person could stay focused on a dot if you also asked them to listen for a bell at the same time. And then they got other grants to study how long it took you to press a button after seeing a flash if you had gotten only two hours of sleep in the last two days. Later they got even *bigger* grants, to see if you could keep track of different instructions being streamed into your ears—some on one side, some on the other. And how long could you *monitor a radar screen* while this was happening?

That last question gives away the game. Because it is a deep and important fact that the vast majority of the research done on "human attention" in the twentieth century was tightly tied to very practical problems concerning the stimulus-and-response capacities of human subjects sitting in front of *machines*.

This wasn't an accident. The primary funders of such work were immensely concerned with human beings as reliable components of big, expensive, intricate *systems*. The military-industrial complex increasingly placed human beings at critical nodes in breathtakingly powerful and costly machines (and *arrays* of machines). For their purposes, what was relevant about "human attention" was the way its parameters established the limits and characteristics of human-machine integration. The science of attention in the twentieth century was a *cybernetic* science, centered on understanding how long humans could stare at screens and dials, how reliably they would click and swipe, how consistently they could track and trigger amid a complicated array of signals. The study of human "vigilance," and the emergence of a quantitative analysis of human attention in information-management terms, all took shape in a set of subdisciplines known as "human factors research." Which is to say, research into the way humans were a "factor" in power plants, giant factories, and, above all, the defense industry.

Is the thing that was scrutinized and quantified in those laboratories "attention"?

Sure. If you like. It certainly *can be TREATED as attention.* And you can measure it, which is nice. Plus, you can track it. Which is also convenient. Indeed, much of the early work in this area centered on making "eye-tracking" systems, which, long before lasers and computers, could be used to monitor the ocular focus of an experimental subject. Were the plot-diagrams of visual trajectories that could be drawn with such systems "attention maps"? Absolutely. And, moreover, they could be used to figure out the "attention value" of magazine ads at different sizes in different positions on the page. Indeed, as early as the 1930s, it was *actual magazine companies who were funding such research,* so that they could more precisely monetize their readers. A very "beta" build of the early attention economy.

But here's the point: When we fret, now, about our "attention span," we are fretting about our attention *in terms given to us by THIS research tradition.* The very idea of an "attention span" in a durational sense was an exact outgrowth of such studies (actually, earlier in the twentieth century the notion of an "attention span" referred not to duration, but to the actual visual field in which a stimulus might appear—it was spatial). Is THAT the attention we are fighting to take back from the frackers?

It would be a hollow victory. Since that kind of attention was literally *defined* as the capacity to "select for a task." It was pre-stressed for cybernetic instrumentalization of the human. Reclaiming it would simply equip us to *select for other tasks.*

But the fullness of human attention is so much more than that! And you know that. If you aren't sure, go take a walk!

So are we saying that modern attention science is *wrong*? Useless? Some kind of *conspiracy*? No, no, and no! Rephrase the question:

Are we saying that if a bunch of Buddhist monks in Nepal had been given, in 1940, the entire research budget of the U.S. military to do scientific research on "attention" for forty years, we would have a very different "science" of human attention?

Yes. We *are* saying that.

How different could the study of "attention" look? *Very* different. Indeed, entire worlds of incredibly deep and serious work went into the study of human attention long before army aviators were strapped into devices to measure their reaction times (as they were gradually deoxygenated—to make it harder . . .).

Take the shocking work of Augustine of Hippo, the fourth-century North African philosopher who wrote a landmark work of autobiography, *The Confessions,* during the late Roman Empire. In chapter eleven of that work, Augustine digs into a searching reflection on the nature of time, and comes to argue that humans are effectively incapable of pure, true, focused attention *because they exist in time.* Forever splayed out across memories of the past, anticipations of the future, and the vicissitudes of each passing moment, humans can *never* genuinely gather themselves into the conditions of sensory and cognitive unity that would be a state of authentic "attention." What did this mean? It meant that our distractibility was a mark of the Fall, and that our inability to give our full attention to anything was a direct result of those fateful events in the Garden of Eden.

The implication? The pursuit of moments of the best kind of attention humans can achieve—moments of authentic contemplation—was understood by Augustine (and many of those who followed him in Christendom) as nothing less than *redemptive.* The pursuit of attention was an effort to *return to God,* from whom we had come.

Way, way, *way* different than trying to shoot down an airplane or spot the pip on a radar screen. And a reminder of the incredible range of ways human attention has been studied, pursued, and understood.

But so what?

Well, this kind of thing actually does matter. After all, when you are pushing for revolutionary change, and you are advocating radical new kinds of understanding, the fact that, by and large, everyone is mostly stuck on one very narrow and specific concept is a genuine problem. And here, because scientific ideas are so technically powerful and authentically compelling, the best practices of well-meaning scientists can actually present very real "blockages" for deep, true, and necessary transformation.

Because there *really is more to it than that.*

Let's review some examples. Not because we are anti-science. But because we feel that it is *genuinely* important for you, a reader of this book, and for all the people who care about Attention Activism, to put aside (now and again) that well-meaning reflex to defer to anyone who identifies as a scientist. Be polite to them! They are great, and we need them in lots of ways! But where attention is concerned, you know every bit as much about the subject as they do. Because human attention is *yours.* Take it seriously yourself, and let's turn to the work of figuring out for ourselves what it is, and what we can do with it.

For starters, then, just about any thoughtful person who has done a little research will be able to call out an instance or two of when the "scientific" propositions of a given moment turned out, later, not only to have been wrong, but to have been wrong in ways that were fundamentally at odds with collective progressive movements. Let's take a particularly monstrous and shockingly recent example. In September of 1950, when a twenty-one-year-old Martin Luther King Jr. enrolled to audit a class on Kant and aesthetics at the University of Pennsylvania, the most prominent member of the anthropology department there was a recent arrival from Harvard, Carleton Stevens Coon—*a vigorous proponent of scientific racism.*

Coon was up to his elbows in highly technical debates about population genetics and biological evolution. He represented and defended, using the very best science of his era, positions that were nakedly racist—and nobody but another scientist of highly specialized training would have been able to argue with him on his own terms. Some did, of course. But much of the argument stalemated for years—even among the experts. Only in retrospect are his ideas widely perceived as noxious, and obviously limited, biased, and fundamentally wrong.

Would young Martin Luther King have been able to "defeat" Professor Coon in a *scientific debate* about the racial inferiority of Black Americans? No. Not in the discourse of "science" of the day. King wasn't a scientist.

But would he have *deferred* to Coon? No way. He would have had to say something along the lines of "I'm not going to argue with you in that language, using those tools; there are a wider set of issues at stake here, and I strongly suspect that, in ways you don't even understand, the whole UNIVERSE of the analysis you are using—the presuppositions, the questions, the framings and definitions—is itself the product of a *history* and a *culture* that totally condition and even, I suspect, *determine* your findings."

King would have been, of course, totally right.

No less compelling, in the annals of science-and-society, is the example of the half century of research on tobacco and human health. Despite there being *plenty* of anecdotal evidence that smoking was bad for human beings (people already called cigarettes "coffin nails" way back in 1900!), it proved amazingly difficult to marshal scientific proofs sufficiently indisputable as to force regulation (and drive successful tort litigation). In this case, historians have done a TON of research to uncover all the ways that interested parties—mainly the tobacco companies themselves!—funded a vast network of scientific researchers who generated *huge* quantities of actual science (peer-reviewed papers in scientific journals, etc.), all of which *obscured* the links between cigarette smoking and morbidity. For the activists who were pushing legislation on tobacco, this kind of science was the *problem*—to be defeated by *other* science (sometimes), as well as by an alternative focus on wider cultural and social norms.

Upshot? Scientific accounts of things are great, and cannot be ignored; but science is always *PART* of culture (it isn't like it comes from outer space! or directly from the mind of God!), and that means the questions that get asked by scientists, as well as the answers they come up with, need to be thought through in the most expansive and critical ways. Especially by activists, who are frequently trying to get everyone to see something *really important* that is also *a little off the radar screen*.

The two examples given above present slightly different versions of the core problem of science's entailment to culture. Racist thinking was *pervasive* in the United States in the 1940s and '50s. Physical anthropology in the period was thus suffused with ways of thinking and feeling that most people in America in the twenty-first century would find repugnant (we hope!). A scientist like Carleton Coon brought his ingrained biases to his research and performed science that confirmed those biases. The tobacco scientists who were deep in the pay of the tobacco industry were certainly raised in a culture that was more generally sympathetic to smoking than our own. But the situation was nevertheless driven by something other than ambient bias: In the 1960s, such researchers had been actively *captured* by a formal strategy of monied interests, who were actually paying (in the form of research grants and prizes) to get the "science" they wanted—in order to protect and extend their financial commitments.

Where contemporary attention research is concerned, one can find examples of both of these distorting problems, general bias and bought-and-paid-for "corporate science." Without wanting to suggest that having a narrow or biased view of human attention is in any way the same (ethically, conceptually) as having a narrow or biased view of human racial identity, it is importantly the case that we live in a world in which just about everyone already thinks that "attention" basically "is" the ability to stay focused on a screen-based task—especially one that is pretty tedious and repetitive.

As it happens, the actual clinical test for disordered attention is, unsurprisingly perhaps, so close to a video game that a number of researchers recently experimented with letting kids play an *actual* video game instead! On the one hand, one might sort of say "Hmm, that makes sense, if a kid cannot keep watching a screen for a long time and hit the space bar when a certain word or letter appears, that kid may have some kind of attention problem." On the other hand, one might step back from this experimental setup and ask, "How the HECK did *this* become the way we think about *human attention?*" Or, to put it another way, "Why is this kid sitting in front of a computer, looking at a screen, and being asked to click away in response to random stimuli? Isn't there a lot more to the universe—and to human attention—than THIS?"

We think those are good questions to ask! And as we have suggested here, we think there are answers. To sum up, and put it simply, the twentieth-century laboratory study of the thing called "attention" was, when one really digs in and investigates it, almost entirely a research project driven by very specific Cold War thinking. The model of the human person from which this research moved was a model of an input-output device, and the laboratory scenarios that were used to examine attention focused almost exclusively on stimulus-and-response scenarios. At issue? The problem at the heart of information-intensive systems where humans had to sit in front of machines (in factories, radar stations, and communication networks).

A century of laboratory studies (funded by the advertising industry and the military-industrial complex) sought empirically to observe, measure, and quantify human attention. What they found, over time, was *a kind of attention that could be measured and understood with the tools they had.* This attention was all about how long a person could keep their eyes on a screen. That is to say, this attention was primarily *visual,* meaning the other senses got more or less lost in the shuffle. It was *durational,* meaning that it could be measured in numbers of seconds and minutes. It was *cybernetic,* meaning it was about the relationship between a person and a machine.

Could "attention" be understood in other ways? For sure! But did the specific cultural preoccupations of the twentieth century canalize the research into a few narrow channels? You bet! And that basic fact goes a long way to explaining how your kid can end up with a clinical diagnosis of "Attention Deficit Hyperactivity Disorder" because of a substandard performance on a game of digital whack-a-mole (we are only sorta kidding—the game the scientists used is actually called "Running Raccoon"; you can look up the paper).

It is perhaps worth remembering how strange this diagnostic scenario would likely look to a Yanomami shaman. What, he might wonder, are those people *testing*? What possible relationship could it have to the WORLD?

Well, that's a good question. In the greensward of Amazonian jungle south of Roraima, *not much*. To the world we have made over the last century in the United States and elsewhere—a world in which our bodies and senses have been remade as appurtenant to screens and keyboards—such a test seems to go to the heart of our humanity. But there is nothing natural or inevitable about that. That is culture. And history. And the science we have reflects both.

Taking back our attention from the human frackers doesn't mean fighting in their ring by their rules. That would be *already to lose*. We don't want to scrap for slightly longer "attention spans," or use their apps to train ourselves to a higher pitch of cybernetic "focus." We want *human* attention, OUR attention—the fullness of that, in all its radical and revolutionary power.

ATTENSITY!

A Manifesto of the Attention Liberation Movement

You are correct: Something is seriously wrong. It has to do with our ATTENTION, our essential ability to give our minds and senses to the world. This precious capacity has been channeled, captured, and commodified by an industry of immense technological and financial power. How? Call it "human fracking."

Human fracking is bad for people, and for politics. It reduces our very beings (and our relationships) to that which can be quantified, bought, and sold. **All this is the triumph of a catastrophic *lie* about what it means to be human.** But deceit and exploitation are never inevitable. To push back, we need more than isolated, individual efforts; what we need is a movement of *collective resistance*.

This movement of attentional liberation exists and has a name: ATTENTION ACTIVISM.

Attention Activism is a fight for justice. This emancipatory uprising takes our apocalyptic present, turns it on its

head, and creates, from the chaos and confusion, new vistas of human flourishing.

Attention Activism is rooted in STUDY—a commitment to diverse forms of teaching and learning centered on attention (what it is, what it can be, what it can do). Attention Activism also requires COALITION-BUILDING— collaboration and solidarity across a range of communities who see attention's essential role in human flourishing. Finally, Attention Activism means the formation of SANCTUARIES—spaces where people can gather, care for each other, experiment with different kinds of attention, and conceive brighter futures.

To discern the revolutionary possibilities of the present, we look to artists, thinkers, and dreamers. To bring those possibilities to *bloom*, we heed the countless Attention Activists who are already out there, devising new (and revising old) ways of giving their minds and senses to each other and the world.

These *attentionauts* and *attentionistas* draw on the wisdom of diverse traditions. Across uncharted terrain, emerging practices of joint attention illuminate new horizons of shared political power. Not only power, but beauty, and grace, too.

This is our movement: the free movement of attention in its fullness, freely shared. We call that transformative goodness *ATTENSITY*. Join us in this heightened and heightening glory—or let us join you!

All this is the triumph of a catastrophic *lie* about what it means to be human.

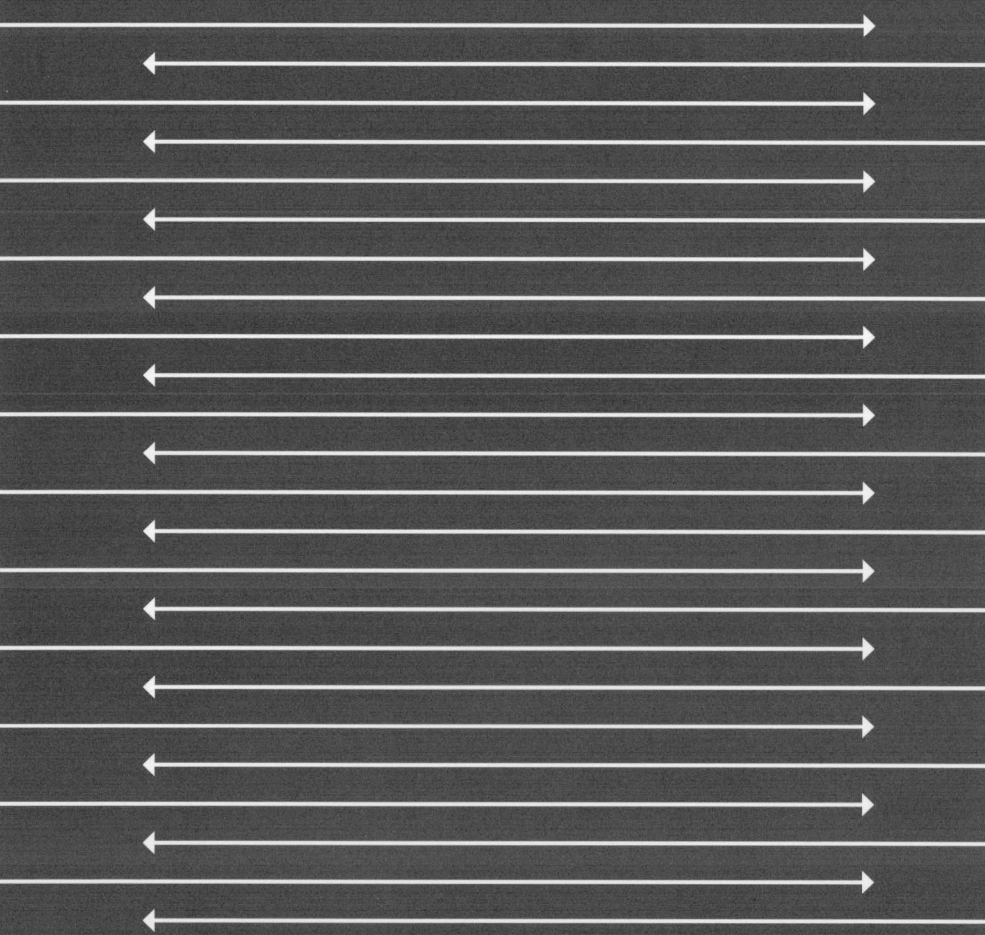

At stake in this other conception of attention? A more dignified, just, and fundamentally TRUE vision of what it means to be human.

n fact, we will go so far as to say this: *At the heart of the attention economy is a LIE.*

It is a lie about what attention is, and how it works, and what it is for. It is the most pervasive and sinister kind of lie—that is, the kind of lie that has been so deeply baked into the way that our world is put together that it becomes invisible. Its disguises? Shapeshifting and familiarity. Where originally the lie was a verifiable claim (something that can be true or false, or in-between, but which, at minimum, merits careful scrutiny and active judgment), now it is a simple taken-for-granted characteristic of our technology, our education, our social lives, and our self-regard. When such a transformation happens—when an idea disappears into plain sight—something else disappears with it: the awareness that there is an *alternative.* MANY alternatives, in fact. A whole universe of them! And since every idea can, as we have seen, shape the world in its image, a whole universe of alternative ideas represents a whole universe of alternative worlds.

These are the worlds that interest us. We have reason to believe that many of them are more conducive to human flourishing than the one we've got. But the first step is to *spot the lie.*

So, what is the lie? It is a claim about the nature of attention. It is the lie that attention "is," as the cognitive neuroscientists put it, "selection for a task." That attention "is" this narrow, specific, measurable characteristic of mental "focus"—a stimulus-and-response feature of the human organism when it is being tested for its capacity to track, trigger, and "vigilate."

Is that really a LIE?

How can it be? There are thousands of highly trained scientists in hundreds of laboratories around the world who have produced countless scientific papers about *exactly this.* They can't all be "wrong"! Well, in the last section we laid out the very particular history that has bequeathed this kind of attention-thinking to us.

To call digital whack-a-mole "human attention" is a little like saying human marriage "is" the "legal framework for the administrative regulation of human reproduction." If you met someone who said that, you would say, "Hmm—I get what you are saying, but that just isn't really right; there's more to it than that." And if you were stuck in a conversation with someone who really just kept INSISTING that was *it* (that marriage was state-bureaucratic-baby-paperwork and that was all there was to it), you would probably feel, very quickly, like you were ready to keep moving around the party. That account of things *just isn't good enough*—isn't true enough to people and their experience, to history and aspiration and life—to satisfy. Everybody is entitled to their belief, and this person is entitled to theirs! But then it is up to them to live with what they think! Good luck on that, since it seems frankly impoverishing.

Our impoverished understanding of attention is like that. Sure, Candy Crush Saga involves "attention." But that is not the only kind of attention in the universe. And it is *certainly* not the most important kind (remember Weil's invocation of *faith*? and *love*?). The problem is that the Candy Crush kind of attention (visual, vigilant, reflexive, cybernetic, quantified) proved IMMENSELY USEFUL to the advertising industry and the military—so useful that it seeped into every other part of our lives. Educators began to think of attention in these terms. So did parents. Even artists. And the computer programmers who shaped the architecture of the digital spaces where more and more people—from the mid-nineties through the aughts and into the present day—spent more and more of their time. Now, when a person says they are trying to "pay better attention," it is more likely than not that they mean *this* kind of attention—the kind that measures their ability to stay with (mostly screen-based) *tasks.*

The lie is that this kind of attention is the only kind of attention. What's the truth? That there are *so many more.* Indeed, the WORLDS of human attention are innumerable—genuinely infinite, and full of infinite promise.

It is our hope, in this short volume, to gesture at a few of them, and, more importantly, to describe the ways that a growing network of human persons joining in conscious resistance to the frackers can come to discover and bring to life their diversity. This is a matter of health, and flourishing, and justice. But it is also a matter of *truth.*

This position—that our issue is one where the actual truth is in confrontation with genuine bad actors—places us in a long genealogy of social (and political) movements like the fight against Big Tobacco or Big Oil. In both those cases, powerful coalitions of industry executives, bought-off scientists, government regulators, and greased-up politicians waged an all-out war on truths that threatened their bottom line: the truth that cigarettes are terrible for our health and for the health of kids, and the truth that carbon dioxide is demonstrably linked to catastrophic levels of planetary warming. These ideas were no secret—they were, in retrospect, astoundingly well-known! Common knowledge, even! But that meant little to the tobacco and oil lobbies, for whom metaphysical categories like truth were annoying abstractions. What moved the needle (in the case of Big Tobacco; the fate of Big Oil remains uncertain) was the ability of a coalition of community groups, advocates, scientists, and crusading litigators to *insist upon the truth,* again and again, and to insist that it mattered—for real! These arguments brought the matter of Big Tobacco's denial back to the site of its original harm: the bodies of the smokers, and, critically, the non-smoking children who were harmed by the noxious, tarry clouds.

So, too, with the frackers. They are harming us. But the lie in our case goes even deeper. This is not simply an argument about whether human fracking is bad for us (although that is an important fact, and a true one); it is a dispute about *what attention is.* And because attention is the way that we enter into the world and shape it, and the way the world enters into and shapes us, the argument about what attention is ends up being, in effect, an argument about our *beings*—about our very humanity.

In this sense, we take up the commitments of those engaged in the long struggles for dignity, such as the suffragists and the civil rights marchers. In their battle for equal treatment under the law, they stood up for a *truth about the nature of people themselves.* These are disputes about human dignity—about the fullness of what people are and what they deserve. Attention Activism stakes a claim for human dignity. We humans are *more* than that which can be quantified and optimized. We deserve a world that is true to our fullness. It is our just inheritance. We therefore demand a world that accommodates the range of human *attentions*—a world, in other words, that reflects the truth of who we are and what we are capable of creating.

This is the highest truth we stand for. And it is *true.* But it is "true" in a way that is distinct from the "truth" of, say, the correlation of secondhand smoke exposure with respiratory infections in children. Our truth cannot really be "proved" by any existing scientific methods. It cannot be "pinned down," exactly. It is too big for that! In fact, it is precisely the nature of this truth—that our humanity cannot be reduced to numbers, or even to language—that makes it impossible to put forward in a way that will "compel assent"! This truth is infinite, because it is the truth of our *infinitude.*

Afraid of truths like that? Fair enough. Worry that under the banner of the "infinite" people have done all kinds of weird and bad and destructive things? Okay, yes—that's true.

But the history of efforts to *avoid* such truths is plenty messy, too. Want only the kinds of truths that can compel assent, in math or logic? You are likely to find yourself in a world you don't like. Because freedom is freedom, and freedom is central to the beings we are—the beings we know we can be. Those beings, *these* beings, need something other than "compelled assent." We need the open space of exploration, the glory of discovery, the experience of understanding.

This has implications for how we move through a day, and how we move as a movement. We must organize around a truth—authentic human freedom—that we *actively seek to understand.* We use the full range of our attention at the same time that we are learning precisely *what that range is,* and *what we can do with what we learn.* We rally around a banner emblazoned with a question mark and an exclamation point—a symbol of our emphatic commitment to an ever-unfolding understanding of who we are and what we are for.

And here's the fun part. YOU help uncover and realize this truth. The more people who join in this work, the more *real* this truth becomes.

ATTENSITY!

A Manifesto of the Attention Liberation Movement

You are correct: Something is seriously wrong. It has to do with our ATTENTION, our essential ability to give our minds and senses to the world. This precious capacity has been channeled, captured, and commodified by an industry of immense technological and financial power. How? Call it "human fracking."

Human fracking is bad for people, and for politics. It reduces our very beings (and our relationships) to that which can be quantified, bought, and sold. All this is the triumph of a catastrophic *lie* about what it means to be human. **But deceit and exploitation are never inevitable.** To push back, we need more than isolated, individual efforts; what we need is a movement of *collective resistance.*

This movement of attentional liberation exists and has a name: ATTENTION ACTIVISM.

Attention Activism is a fight for justice. This emancipatory uprising takes our apocalyptic present, turns it on its

head, and creates, from the chaos and confusion, new vistas of human flourishing.

Attention Activism is rooted in STUDY—a commitment to diverse forms of teaching and learning centered on attention (what it is, what it can be, what it can do). Attention Activism also requires COALITION-BUILDING— collaboration and solidarity across a range of communities who see attention's essential role in human flourishing. Finally, Attention Activism means the formation of SANCTUARIES—spaces where people can gather, care for each other, experiment with different kinds of attention, and conceive brighter futures.

To discern the revolutionary possibilities of the present, we look to artists, thinkers, and dreamers. To bring those possibilities to *bloom*, we heed the countless Attention Activists who are already out there, devising new (and revising old) ways of giving their minds and senses to each other and the world.

These *attentionauts* and *attentionistas* draw on the wisdom of diverse traditions. Across uncharted terrain, emerging practices of joint attention illuminate new horizons of shared political power. Not only power, but beauty, and grace, too.

This is our movement: the free movement of attention in its fullness, freely shared. We call that transformative goodness *ATTENSITY*. Join us in this heightened and heightening glory—or let us join you!

But deceit and exploitation are never inevitable.

All revolutions begin with the same insight: *It doesn't have to be this way.*

n a groundbreaking book published in 1978, the American historical sociologist Barrington Moore Jr. went after a fundamental question: Why do large numbers of people so often put up with horrible and unfair circumstances for so long?

It was the question at stake in his simple title: *Injustice.* Taking up searing test cases (the "untouchables" of South Asia, brutalized industrial laborers at the turn of the twentieth century), Moore drilled down, trying to understand the conditions that make people submit to situations that seem manifestly abhorrent—and not obviously necessary. His aim? To identify the dynamics that lead to revolutionary change. In a powerful passage that combines psychological acuity with the hard-won wisdom of the archive, Moore summed up the crux of the issue:

People are evidently inclined to grant legitimacy to anything that is or seems inevitable no matter how painful it may be. Otherwise the pain might be intolerable. The conquest of this sense of inevitability is essential to the development of politically effective moral outrage. For this to happen, people must perceive and define their situation as the consequence of human injustice: a situation they need not, cannot, and ought not to endure.

These words capture the essence of our current condition. We are *in pain,* suffering from a pandemic of loneliness and alienation and suspicion, trapped in impotent cycles of submission to . . . "ourselves" (to what seem to be our own reflexes and appetites, to our own enflamed anxieties and deformed aspirations), but in ways we know are, somehow, *not who we actually are.* This is the existential phenomenology of creatures who are literally being bio-hacked at a societal scale—and it feels *terrible.*

At the same time, Moore has spotted that there is something *even more terrible:* The painful awakening to a recognition that *it doesn't have to be this way!* Why does this hurt so much? Because this haunting insight opens onto a new anguish: We *COULD* have it otherwise! Which means that we have been suffering *needlessly.* And this is super hard to confront. Easier to nest in the learned helplessness of victimization, the psychological response of defensive resignation.

But every true movement of revolutionary change must confront and overcome this all-too-human logic of whipped-dog acquiescence. What we need is *politically effective moral outrage.*

How do we get there? We focus on a simple truth: The wholesale fracking of human beings for the money-value of their humanity (their interest, their curiosity, their sense of freedom, their capacity to care—in short, their *attention,* understood in its full richness and complexity) is not "natural." It is not "inevitable." It is the historically anomalous result of *human actions*—which is to say, the result of the decisions and commitments and planning and programming of specific persons; and the *failures to act* of others. Greed and heedlessness have been the basic drivers, both of the actors (Big Tech) and the do-nothings (the regulators and legislators). It is the aim of Attention Activism to speak loud and clear: **It does not have to be this way!** And you have every right to be OUTRAGED that things have taken this shape!

What we have here is the pure stuff of Barrington Moore's revolutionary condition: a situation that we need not, cannot, and *ought not* to endure. Once we all see that, once we all feel it, together; once we have all conquered the too-easy sense that a vampiric media economy feeding on the lifeblood of our humanity is "inevitable," *THEN* we will sweep ourselves to a new world—indeed, coming to that shared awareness will itself BE the core of the revolution, since this one does not require killing kings or over-throwing a government.

Barrington Moore has some things to say about activists, too, and their role in the dynamic process of radical change: "They do the hard work of undermining the old sense of inevitability." A key strategy in this regard? *Remind people of the ways things have changed in the past.* Because large-scale social change has actually happened, again and again. From the most majestic and improbable (the Civil Rights Movement in the United States miraculously effected the "transformation of the consciousness of millions of Americans") to the most ghastly and calamitous (Haiti has largely ceased to function as a working polity within the last decade). Telling stories about change in the past helps us see that change is always ahead, and helps shatter the inertial complacency that consistently makes *now* feel like it has a hold on forever.

So let's remind ourselves of a really significant change in the life of the United States that has happened within the lifetime of at least some of the readers of this book: the rise of an "exercise culture" across the last half century. It is hard to overemphasize how new and how broad this change has really been. It will shock any American under the age of fifty to learn that, once upon a time, nobody in the United States "went running." Indeed, you couldn't go to the store and buy a pair of "running shoes" in, say, 1965— because such a thing *did not exist.* Sure, Phil Knight was somewhere out in Oregon, driving to track meets with a lightweight waffle-rubber-sole shoe in the back of his station wagon. And yes, there were spiked sneakers for Olympic athletes. But, basically, a pair of Chuck Taylor "All Stars" (canvas, flat, originally designed circa 1915) were pretty much still your best option if you wanted to go for a jog.

But you *didn't* want to "go for a jog." Because "jogging" *wasn't a thing.* Okay, maybe you were in training as a college athlete or something. Or perhaps you were in boot camp and being punished at the bugle call of reveille. But otherwise, normal people did not come home from work, throw on a pair of shorts, and "go out for a run." They didn't "go to the gym," either. Unless they happened to be a powerlifter or a prizefighter. Because there were no "gyms" that welcomed ordinary go-to-work-on-Monday-morning professional adults. "Gym" was for high schoolers. Grown-ups of the middle class might go to a "club," and they might play tennis or golf. But they certainly did not have a "trainer"—and taking a walk was understood to be "exercise."

What happened?

What happened was really a kind of "revolution"—though not a French Revolution–style revolution. Something more like a quiet, gradual, general, and very real *transformation in culture.* And it was driven by a host of different factors, including actual public policies (the U.S. Army fretted about the conditioning of its new recruits, and helped encourage a fresh emphasis on "physical education" in schools), popular authors and books (James F. Fixx's 1977 bestseller, *The Complete Book of Running,* did more to spur health-buff joggers than any other single event or phenomenon), and eccentric impresarios and showmen (like the irrepressible fitness entrepreneur Arthur Jones, who developed both the BowFlex workout station and the Nautilus line of gyms and exercise machines—and who also spent a fair bit of time trying to fatten his pet saltwater crocodile to world-record proportions). This was a do-gooder revolution of advocates and nonprofit programs on the one hand, and *also* a revolution driven by bottom-line investments and commercial success. The breadth and scale of the transformation can be read in some actual statistics: In 1960, less than a quarter of Americans claimed to exercise "regularly"; by 1987, that percentage had nearly tripled. In 1972, 1.7 million Americans belonged to a fitness club of some kind. Fast-forward to 2006 and that number is a shocking 42.7 million—that is a growth of 2,500 percent! This has meant large returns for a number of market players: fees on club memberships hovered around $227 million in 1972; by 2023 that revenue topped $33 *billion.*

Why review all this? Because the remarkable growth of the fitness sector over the last fifty years, and the wider cultural transformation of which it is a part, offer a powerful image of what we believe lies ahead for Attention Activism. We are convinced that historians of the future will look back on 2010 with the kind of surprise we feel when we reckon with the idea that nobody was "jogging" in 1960—only they will be saying, "Yes, it's true, back in 2010 there was, shockingly, no meaningful cultural awareness concerning 'attentional well-being'; people didn't really *exercise* their attention in any formal way, and there was no general, collective commitment to the idea that keeping your attention 'in shape' was an essential aspect of human flourishing." Listeners will shake their heads in disbelief! How could so many people have totally overlooked something so completely basic?

We believe a comparable change is possible—indeed, that it is on the horizon! And here's our wager: Attentional wellness—the training and conditioning and maintenance of one's diverse attentional capacities—will, in 2040, be understood as a *basic good;* simply pragmatic, on the one hand (because one is more effective in the world when one's attention is healthy and well), but also more than merely "useful," since attentional wellness is also *beautiful* (it brings joy, and makes for richer lives, deeper intimacies, and a better world).

To see the transformative potential of the present moment—to understand *what we can change in ourselves* and *how we can resist what has been done to us;* to get our attention *off the couch* and onto the pavement—we must bear in mind this simple fact: ATTEN-TION IS *LEARNED.* Always. You were taught to listen the way that you do. To look, too. And to feel.

Exactly *how*—that is, exactly who taught you—is a muddier matter. It's no less loaded than the matter of *how you came to be who you are.* But in broad strokes, the answer involves some combination of the attentional habits of the people who raised you, the schooling you received, and the informational ecology that teaches you, every day, what kinds of sights and sounds to listen and look for and how. Most of this learning goes on without our knowing it. We're always learning how to attend. You're doing it right now.

So, on this, there's bad news and good news. The bad news is that the attention frackers have gotten really, really good at sneakily teaching us how to attend. They've also gotten really, *really* good at getting us to maximize the amount of time we spend in *their* classroom—which, thanks to some clever tech, is small enough that anyone capable of using the toilet alone can hold it in the palm of one hand. Think of it as a glowstick-bedecked rodeo clown who bursts into your second-grade art class, mounts the table (where you were all quietly making paper doilies or something), and starts shouting, *STOP! LOOK AT ME! RIGHT NOW! I'M GOING TO TEACH YOU HOW TO SEE THINGS—EVERYTHING—MY WAY!* And you do—every day, most of the day, for the rest of your life. Except the clown sends you work emails. And porn! And stuff that is almost like porn, but not quite (even better!). And an ad for a new kind of lip balm. And it all fits in your pocket.

The attention frackers have monopolized the once-sacrosanct business of attention formation, and they have done it by dominating, with oodles of money and unimaginable computing power, a deck-of-cards-sized patch of space approximately eight inches from your face. This results in a population that is systematically over-educated to one very specific kind of attention.

But here's the good news: Everything that is learned can be un-learned, and relearned. Can be *trained* (cue *Rocky* theme music)! And that is the task before us. It is, in a way, to make ourselves will-fully *non-commodifiable* in our attentional lives.

Put differently, this requires that we learn (and PRACTICE) the kinds of attention that are *not* engineered to maximize digital advertising revenue. And that is a whole lot of them. In fact, the vast majority of the range of human attention falls into this cate-gory. If the task-oriented, durationally optimized attention that the frackers encourage is *one* "kind," you can think of the rest—those kinds of attention that we call "radical" attention—as being, well, *infinity-minus-one*. Which is to say, infinite. So we have plenty of tools in our kit.

But we are mostly not accustomed to thinking in this more ex-pansive way. And that's because we've mostly been trained (by screen-work, by screen-school, by screen-play, and above all by the human frackers themselves) to think of attention as a singular thing. As our ability to poke this stimulus. Every time it flashes or pings. Our ability to stay with the stream of notifications. Our im-mersive swipe-o-philia.

But it is essential to think *otherwise*. Because that way of think-ing is literally killing people. It is delivering us to the algorithms, which press a sickly trickle of cash money out of our actual lives (to flow into the bank accounts of greedy profiteers). They press the money out of our eyes. Out of our friends. Out of what we love. Worse, out of our need for love—and out of our *ability* to love.

It's *okay* to be angry.

As long as you also have hope.

ATTENSITY!

A Manifesto of the Attention Liberation Movement

You are correct: Something is seriously wrong. It has to do with our ATTENTION, our essential ability to give our minds and senses to the world. This precious capacity has been channeled, captured, and commodified by an industry of immense technological and financial power. How? Call it "human fracking."

Human fracking is bad for people, and for politics. It reduces our very beings (and our relationships) to that which can be quantified, bought, and sold. All this is the triumph of a catastrophic *lie* about what it means to be human. But deceit and exploitation are never inevitable. **To push back, we need more than isolated, individual efforts; what we need is a movement of *collective resistance*.**

This movement of attentional liberation exists and has a name: ATTENTION ACTIVISM.

Attention Activism is a fight for justice. This emancipatory uprising takes our apocalyptic present, turns it on its

head, and creates, from the chaos and confusion, new vistas of human flourishing.

Attention Activism is rooted in STUDY—a commitment to diverse forms of teaching and learning centered on attention (what it is, what it can be, what it can do). Attention Activism also requires COALITION-BUILDING— collaboration and solidarity across a range of communities who see attention's essential role in human flourishing. Finally, Attention Activism means the formation of SANCTUARIES—spaces where people can gather, care for each other, experiment with different kinds of attention, and conceive brighter futures.

To discern the revolutionary possibilities of the present, we look to artists, thinkers, and dreamers. To bring those possibilities to *bloom*, we heed the countless Attention Activists who are already out there, devising new (and revising old) ways of giving their minds and senses to each other and the world.

These *attentionauts* and *attentionistas* draw on the wisdom of diverse traditions. Across uncharted terrain, emerging practices of joint attention illuminate new horizons of shared political power. Not only power, but beauty, and grace, too.

This is our movement: the free movement of attention in its fullness, freely shared. We call that transformative goodness *ATTENSITY*. Join us in this heightened and heightening glory—or let us join you!

To push back, we need more than isolated, individual efforts; what we need is a movement of *collective resistance.*

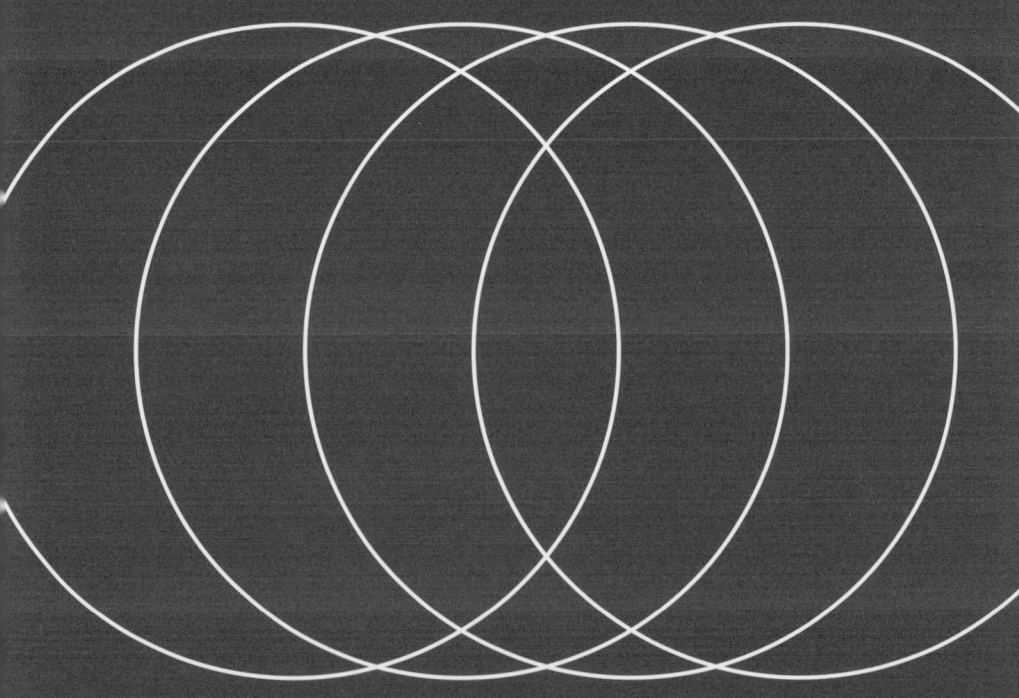

Escape from these conditions is not a matter of individual willpower. We need new forms of collective action and solidarity to confront this new kind of exploitation. History offers examples.

P romises of silver bullets abound, but none will suffice, because there are no quick fixes to a challenge at this scale. No app is going to save us, because there is no "tech fix." No drug is going to heal us, because there's no "Big Pharma fix," either. And don't hold your breath for a *regulatory* fix, because, in the end, no law is going to be able to protect us . . . enough. Self-help regimens and yoga retreats are all well and good, but they are never going to be enough. So go ahead and practice mindfulness, or support sane legislation to fight back against "dark patterns," or try a digital detox. Take your kid to the zoo, for that matter, and try leaving the phones at home. But ultimately the issue at hand isn't going to be solved at the level of individual actions. The problem isn't *personal.* It is *structural* and *systemic.* Indeed, no larger or more complex "system" has ever been achieved by human beings.

And in that context, only some radical new form of collective identity and collective *action,* some truly new social formation, can possibly provide what is needed. What we need is a *revolution* in how we conceive HUMAN ATTENTION and its role in our lives—as individuals, and in communities. To get there, we need a *new kind of activism.* An activism of joint action as we resist and redress the relentless commodification of human attention—the stuff of our very selves.

Are there precedents for a collective effort at this scale? There are. In the nineteenth century, the harnessing of steam power and the elaboration of the "factory system" generated unprecedented possibilities for human exploitation. Up went gigantic new hives for continuous work. Penurious wage labor effectively displaced even chattel slavery as the new maximally efficient, hugely profitable, and utterly amoral nexus of machines, law, and greed. This was the Industrial Revolution.

Facing these impossible conditions, several generations of

social activists, organizers, theorists, and front-line agitators struck back, organizing new kinds of social solidarity to resist these novel forces of unprecedented human power, and the dehumanizing exploitation they engendered (alongside the irreversible and welcome innovations they allowed). The labor movement and trade unions emerged as new forms of collective social action, and these powerful constellations of cooperative thought and practice genuinely transformed the world of the "satanic mills"—and the political life of the last century across much of the world.

We believe that the early twenty-first century presents a strikingly parallel scenario. A new series of technological innovations (massive information storage and processing systems, highly visual and interactive and intimate digital prosthetics, the pervasive financialization of social infrastructure, which has been largely translated into virtual forms) have together created unprecedented possibilities for dehumanizing exploitation—in the form of *human attentional fracking*. Our desire for connection to others; our hunger to see and be seen; our native curiosity and tendency to *care*— all these aspects of our most essential natures have been tapped as access points from which money can be siphoned into the pockets of private equity financiers.

Which is to say, the final frontier for transnational capital would seem to be a precious resource buried inside each of us: our *attention*. The latest global land-grab is a scramble to drive freehold stakes into the very stuff of our consciousness. What is wanted— and what can now be had, everywhere and all the time—is real estate in our senses and in our minds. And that real estate is being seized as never before.

To counter this unholy land-grab, we must look to the tactics of the nineteenth-century laborers, who recognized in their *shared subjugation* the possibility of *shared agency*. We, too, can form communities of solidarity committed to the protection of the very thing that the satanic mills in our pockets take from us: our attention.

Indeed, we must! Any lesser solution—Big Pharma Band-Aids, screen control gadgets, neoliberal wellness practices—will only perpetuate the conditions on which the attention frackers traffic: dependency of one sort or another, the dissolution of our communities, the evacuation of our selves.

What we need is a movement: a COLLECTIVE RESPONSE. One that is commensurate with the scale of the threat. It must be broad (spanning demographics and interest groups), and it must be deep (built upon relationships of care and trust).

Consider, as another instructive example, the diverse set of stakeholders who formed the modern environmental movement between 1965 and 1975. The social forces that led to the establishment in 1970 of the Environmental Protection Agency were driven by youth culture, in part, but the key transformation was the bringing on board of farmers and hunters, blue-collar labor and white-shoe lawyers. During this time, the physical environment came to be understood as a shared problem worthy of collective action. Can we make the protection of the psycho-sensory environment similarly urgent and compelling?

We can. It is a problem at comparable scale and of parallel form. We need a new "environmental movement" to care for the *inner* environment of our minds, capacities, and relations. It isn't less important than the outer environment, and, as we have asserted above, our ability to protect the latter will ultimately depend on our ability to protect the former.

So we need to fight back against the frackers *together*. Because it is going to require buy-in from many communities, and a clear sense of our relationship (as subjects of psychic, cognitive, and emotional *extraction*) to the forces that we are organizing to defeat.

The collective that has authored the book in your hands thinks of itself as exactly the seed of such an alliance of resisters. Working together since 2019 (officially), but tracing roots back more than a decade before that, the Friends of Attention are committed to building a movement grounded in *calling out* the untenability and injustice of our current condition. That means a movement driven by a deep and old faith: Important truths, seen clearly and shared, possess catalytic power; as people see and understand urgent things together, nothing can reverse the transformative force of collective outrage.

We have staked ourselves on that vision. It is critical, and it is affirmative.

But is it *correct*? How can people come together to change things?

These are questions about "movements" and the place of movement energies in democratic politics. In thinking about these issues, we've been led by the recent work of the political philosopher Deva R. Woodly. In her powerful study *Reckoning* (2021), Woodly offers a theoretical account of what movements are and how they work. Drawing on the German sociologist Max Weber, Woodly identifies a central risk that perpetually haunts modern democracies as they stabilize the basic architecture of administrative governance and gain wealth. For Weber, each aspect of that evolution, permitted to run its natural course, inexorably compromises the core of democratic self-governance: Administrative structures gradually harden into the iron cage of Kafkaesque bureaucracy; and the sweet emoluments of increasing wealth inevitably install, eventually, a more or less depraved oligarchy. The result of letting democracy just rumble on? What you will get, sooner or later, is, as Woodly puts it, "dehumanization, expropriation, and stagnation."

Sound familiar? It should. The politics that emerges from such conditions is what Woodly calls "the politics of despair," which characterizes a citizenry that ceases to believe in its own political agency. A perception emerges that the system is "rigged," that, as she puts it, the people in charge are essentially different from the actual people, and are unreachable and unresponsive. Voting comes to seem like a pointless charade. Catastrophism and a grimly pessimistic "realism" prevail. Cynics strut like sages. Meanwhile, structural inequality increases, and the people are ever more profoundly underserved by a political system that progressively feels like an irrelevant and alien imposition.

This account of the "politics of despair" has a shocking and unsettling immediacy. We assume many, and quite possibly most of our readers, feel to some significant degree "seen" in this damning indictment of democracy lapsing into spiritual and practical crisis. And it is precisely such conditions that give rise to, and *require,* in Woodly's account, social movements. Indeed, that is what social movements *are,* as she sees it—a mechanism for confronting the "politics of despair." Her metaphor is rousing: Foresters use *controlled fire* to clear the constricting undergrowth that can imperil woodland ecologies; the term for lighting, tending, and steering such rejuvenating conflagrations is "swailing." Social movements are best understood as a way of *swailing* unhealthy democracies. It's the perfect image: careful fires that aim to make positive change by burning through the bad stuff—to give new life.

We believe in this vision of what a movement can be. And we agree, too, with the way Woodly argues for the deeper significance of movement politics. After all, it is always possible to address the need for political change by going right at the *actual structures of politics*: One can write a letter to a congressional representative, or even run for office, or otherwise try to engage democratically via the available mechanisms of democratic participation. Even when those approaches feel dysfunctional, we cannot merely abandon them—since democratic self-governance will always ultimately depend on precisely such modes of representation.

The Friends of Attention support such efforts, but in confronting the politics of despair, we need something more. We need to remind ourselves, and remind each other, that we—the people—are actually *the foundation of everything*. The way we live and think, what we want and do—this is where democracy begins, and it is also the final court of appeal for every democratic claim. For this reason, Woodly sees social movements as the key to what she calls the "repoliticization of public life." As she puts it:

> They allow us to enact citizenship, not only through performing duties, but also by authoring new understandings, priorities, and even governing institutions. Unlike other forms of participation, which can also teach valuable civic skills, social movements show us how to make change. Even if we do not immediately change policy or restructure institutions, we change our ideas, we change our minds, we change our associations, we change public understandings, and we change the scope of political possibility.

THIS IS OUR WORK. To build a world where people can flourish in the face of the human frackers requires a *movement*.

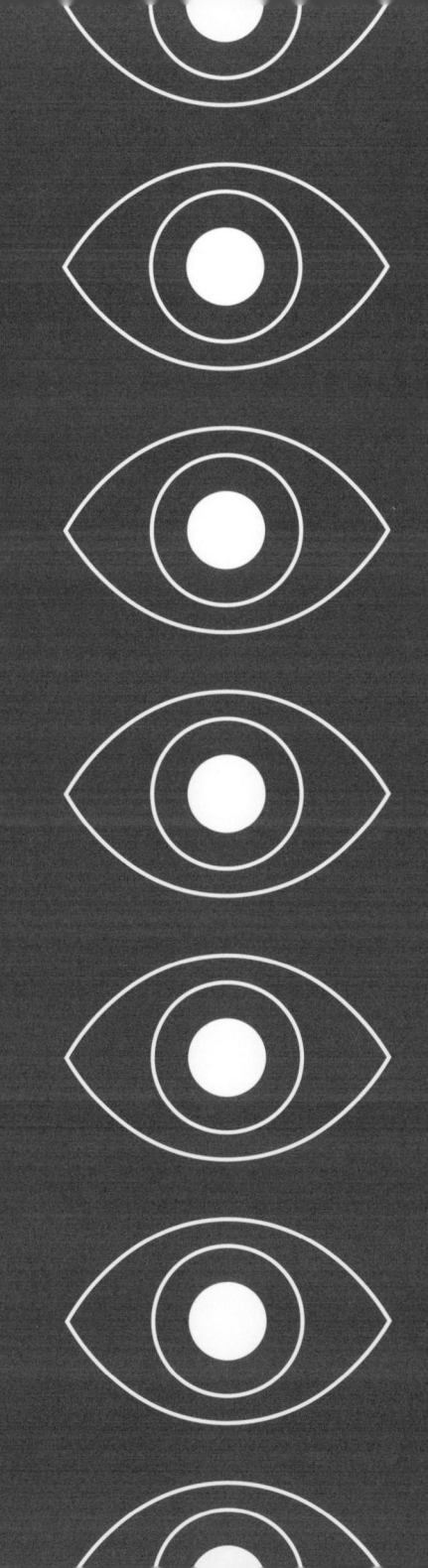

ATTENSITY!

A Manifesto of the Attention Liberation Movement

You are correct: Something is seriously wrong. It has to do with our ATTENTION, our essential ability to give our minds and senses to the world. This precious capacity has been channeled, captured, and commodified by an industry of immense technological and financial power. How? Call it "human fracking."

Human fracking is bad for people, and for politics. It reduces our very beings (and our relationships) to that which can be quantified, bought, and sold. All this is the triumph of a catastrophic *lie* about what it means to be human. But deceit and exploitation are never inevitable. To push back, we need more than isolated, individual efforts; what we need is a movement of collective resistance.

This movement of attentional liberation exists and has a name: ATTENTION ACTIVISM.

Attention Activism is a fight for justice. This emancipatory uprising takes our apocalyptic present, turns it on its

head, and creates, from the chaos and confusion, new vistas of human flourishing.

Attention Activism is rooted in STUDY—a commitment to diverse forms of teaching and learning centered on attention (what it is, what it can be, what it can do). Attention Activism also requires COALITION-BUILDING— collaboration and solidarity across a range of communities who see attention's essential role in human flourishing. Finally, Attention Activism means the formation of SANCTUARIES—spaces where people can gather, care for each other, experiment with different kinds of attention, and conceive brighter futures.

To discern the revolutionary possibilities of the present, we look to artists, thinkers, and dreamers. To bring those possibilities to *bloom*, we heed the countless Attention Activists who are already out there, devising new (and revising old) ways of giving their minds and senses to each other and the world.

These *attentionauts* and *attentionistas* draw on the wisdom of diverse traditions. Across uncharted terrain, emerging practices of joint attention illuminate new horizons of shared political power: Not only power, but beauty, and grace, too.

This is our movement: the free movement of attention in its fullness, freely shared. We call that transformative goodness *ATTENSITY*. Join us in this heightened and heightening glory—or let us join you!

This movement of
attentional liberation
exists and has a name:
ATTENTION
ACTIVISM.

A new form of activism
is on the rise.
And it needs you.

When you read a tag like "environmental activist" in someone's bio, you have a basic sense of what is meant. Sure, there are a lot of different things that person might actually *do* (anything from working full-time for an NGO trying to preserve the Brazilian rainforest, to volunteering an hour a week at a compost collection depot in Oakland), but you know what is being *said*: "I'm somebody who cares about the environment! I think of myself as somebody engaged with others in a big, complicated, and important project that involves trying to make the world a better place—or at least stop it from sucking worse and worse!"

It's important to point out that the very idea of being an "environmental activist" is actually pretty new. Nobody went around identifying that way in 1870—and, for that matter, such a designation was almighty rare even in the 1950s. That doesn't mean nobody cared about the thing we call "the environment." Rather, it took time, and quite a lot of work, to give both those words, "environment" and "activist," the meanings that feel natural and intuitive to us now. There is fascinating history in both those stories: The amazing dynamics by which our surroundings, both natural and civic/urban, came to be understood as complex systems needing collective care and management and protection; the awakening of forms of citizen solidarity and issue-engagement that we consider "activism."

Let's skip all that for now. There are other books you can pick up if you want to read those stories. What's relevant for us, here, is examining the birth of "environmental activism" for lessons as we work to bring forth this new and urgently needed movement of ATTENTION ACTIVISM. Just as early conservationists, sportsmen, naturalists, and urban reformers, working for decades in different parts of the world, had to give shape and immediacy and even a kind of charisma to the notion of the "environment" (linking Romantic ideas of the wilderness to medical ideas about health and well-being, to scientific ideas about chemical and biological interconnectedness), so, too, it falls to those of us fighting the attention frackers to work together—and, when necessary, apart—to raise a thick, true, and *instantly compelling* conception of attention to collective consciousness.

To be sure, attention is already *very much a "thing"* within contemporary culture. People worry about their "attention spans." Schools fret that there's something wrong with the kids, who seem unable to "pay attention." Everyone knows, pretty much, or at least *thinks* they know, the extent to which social media companies and the data barons of the internet monetize our attention.

And all of this is very good. It is part of what indicates just how ready we are for the catalytic emergence of a broad-based coalition dedicated to the protection, cultivation, and even, when necessary, the loving *regeneration* of human attention.

But to get there, we need still more work by everyone who cares: work that dramatizes the richness and profit-defying glory of human attentional ecosystems; work calling out the heedless despoilers of our attentional commons; work standing together to give face and voice and mass to a movement capable of facing down some of the most powerful and well-endowed interests in the world.

Human fracking is a systemic harm. It operates on multiple scales simultaneously. So it stands to reason that any resistance worth its salt will have to move nimbly from the realms of the intimate through the social, the intellectual, the political, the legal, and the technological, to the infrastructural—and back again. Each zone requires a different kind of activism. All of these ways of engaging the problem call on different skill sets and sensibilities. All of them are necessary. This is why we need *all hands on deck*: teachers and artists and builders and technologists and spiritual leaders and lawyers and parents and organizers and scholars and long-haul truckers and meditators and mediators and family and friends. Everybody—*everybody!*—has something to offer.

If you think that could be you, then by all means keep reading. You may find that you are *already* an Attention Activist, and did not know it. Indeed, we suspect that you are.

ATTENSITY!

A Manifesto of the Attention Liberation Movement

You are correct: Something is seriously wrong. It has to do with our ATTENTION, our essential ability to give our minds and senses to the world. This precious capacity has been channeled, captured, and commodified by an industry of immense technological and financial power. How? Call it "human fracking."

Human fracking is bad for people, and for politics. It reduces our very beings (and our relationships) to that which can be quantified, bought, and sold. All this is the triumph of a catastrophic *lie* about what it means to be human. But deceit and exploitation are never inevitable. To push back, we need more than isolated, individual efforts; what we need is a movement of collective resistance.

This movement of attentional liberation exists and has a name: ATTENTION ACTIVISM.

Attention Activism is a fight for justice. This emancipatory uprising takes our apocalyptic present, turns it on its

head, and creates, from the chaos and confusion, new vistas of human flourishing.

Attention Activism is rooted in STUDY—a commitment to diverse forms of teaching and learning centered on attention (what it is, what it can be, what it can do). Attention Activism also requires COALITION-BUILDING—collaboration and solidarity across a range of communities who see attention's essential role in human flourishing. Finally, Attention Activism means the formation of SANCTUARIES—spaces where people can gather, care for each other, experiment with different kinds of attention, and conceive brighter futures.

To discern the revolutionary possibilities of the present, we look to artists, thinkers, and dreamers. To bring those possibilities to *bloom*, we heed the countless Attention Activists who are already out there, devising new (and revising old) ways of giving their minds and senses to each other and the world.

These *attentionauts* and *attentionistas* draw on the wisdom of diverse traditions. Across uncharted terrain, emerging practices of joint attention illuminate new horizons of shared political power. Not only power, but beauty, and grace, too.

This is our movement: the free movement of attention in its fullness, freely shared. We call that transformative goodness *ATTENSITY*. Join us in this heightened and heightening glory—or let us join you!

Attention Activism
is a fight for justice.

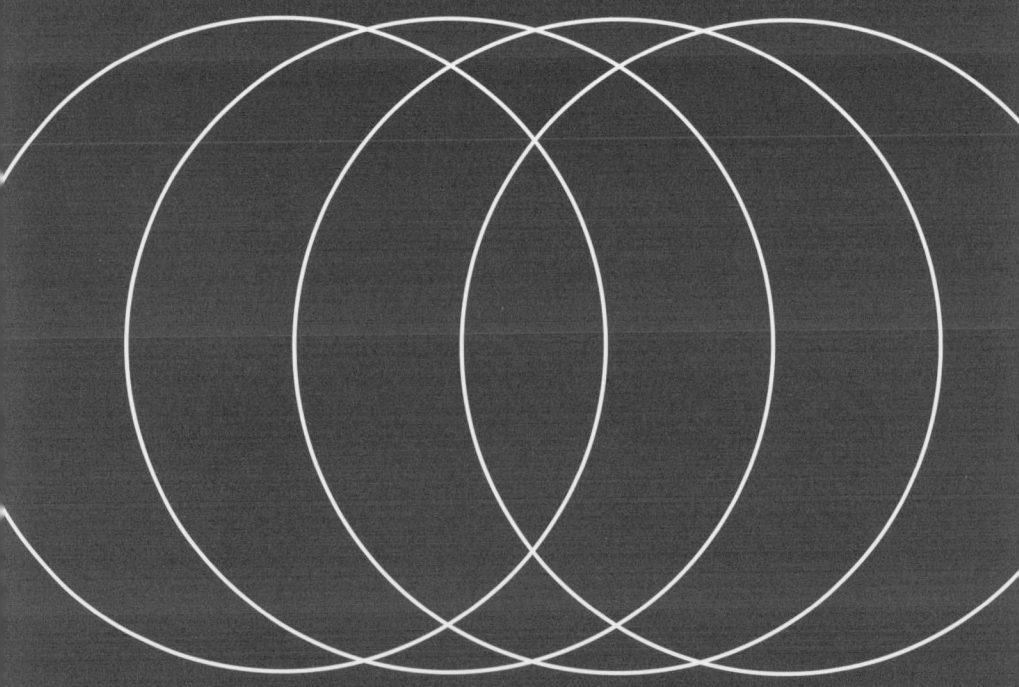

Quantification is the first step toward monetization. And so we refuse the datafication of attention, in order to insist on a fuller vision of human existence.

A s Attention Activists, we commit our hearts and minds to the pursuit of two distinct kinds of justice. The first resists the exploitative and often deceptive extraction of money from our eyes and minds. The second insists that our attention, and our humanity, far outstrip the meager metrics of money-value. These are two very different visions of justice. Both are necessary. We want companies held accountable for unethical practices, "dark pattern" interface tricks, and manipulative design features. That's basic justice. But we want more than that. We *also* want the kind of justice that makes life worth living. Let's take a moment with these different visions, so that we can keep each in its place.

Our first kind of justice will be familiar to many readers of this book. The notion of seeking redress for an unjust loss sits comfortably within the framework of contemporary law, where justice is treated, in its basic form, as a kind of accounting. Recall the image of Lady Justice, bedecked with symbols beside the courthouse door: She stands with a prominent sash about her eyes (representing the impartiality of judgment), a sword in her left hand (the menace of punishment), and a scale in her right. The scale stands for balance. It represents the return of that which has been taken away. It is by the logic of the scale that we as Attention Activists demand the right to reclaim our attention from the frackers.

But let's stay another minute with these scales. For scales measure one property: "weight." Scales imply that this is the relevant quality, and that this property can be measured and compared in terms of quantity. That things of like weight-quantity are, at least in that way, equivalent. What the scale assumes, in other words, is that there exists a measure by which one thing can be found exchange-equal to another thing. This is, of course, only sometimes true.

Which brings us to our second call for justice. This call insists that our attention will always be greater than whichever of its properties we can measure. What this means, therefore, is that to exchange our attention "like" money or "for" money *will always be attended by loss.* Because this thing we call attention is vaster, deeper, and far harder to plot on a balance sheet than the metrics of "time on device" (TOD) that the human frackers leverage.

This is the heart of the matter. We occupy a peculiar historical moment in which what counts as *real* is, according to the dominant ways of thinking, mostly limited to *that which can be counted.* Quantified. Expressed as numbers. And it is no coincidence that this ability to express a thing in terms of numerical value is the necessary first step toward turning that thing into money. This is troubling. It is a logic which holds that everything which exists can be turned into money and sold. Indeed, it is a view of the world which holds that everything is in effect *already* money—even if some part of it is waiting to be monetized.

So when we insist that attention is something more than the movement of the eyes, or a flash or electrical activity in the brain, or several seconds of uninterrupted focus, a different kind of justice is at work. It is not a matter of *who* runs the world, but a matter of *what the world is.* Is the world made of money-value? Or is it made of something else? Something that cannot and will not be accounted for in money terms alone?

Let us plant our flag: We refuse the totalizing and impoverishing logic of money-value. We propose alternatives to the *datafication of being.* We insist that there are other ways of keeping track of existence. We insist, what's more, that some things *cannot be "known"*—at least, not immediately. Not completely. Perhaps not ever. That there is only so much that can be measured. That the stuff which makes life worth living, and worth sharing with others, is *literally immeasurable.* That to ignore this stuff—to deny its reality—is an injustice that denies our very humanity.

As Attention Activists, we are committed, then, to what the legal scholar Tim Wu calls a "human reclamation project." It is a project that actually seeks to protect our *non-inhumanity* by asserting, over and over, that we cannot be reduced to mere salable data. It is a justice that refuses to see the world by the terms in which it has been presented to us: as an exhaustive (and exhausting) inventory of solitary, mute objects waiting to be counted, priced, and sold. In the face of the relentless, blinding quantification of being, we close our eyes, and *listen,* instead, with close attention to those around us.

And here we return to the image of Lady Justice: The justice we call for is not the justice of the scale, nor the justice of the sword, but the *justice of the sash.*

Consider the sash. A flimsy thing. Not an obvious instrument of sabotage (a wrench comes more readily to mind). And given its traditional function as a blindfold, the sash may even appear to contradict our commitment to the freedom of perception. But the sash, like every tool, is mutable. It begs to be played with. And nothing could be simpler! With the sash about your eyes (your gaze thereby *refused* to the eyeball merchants), mark how your hearing grows more acute. Feel the rumble of the ground beneath your feet. Now remove it. Examine the back of your hand (that clichéd site of the *always-already-known*). Look weird? Unfamiliar? Maybe even . . . beautiful? The sash stands for our ability to withhold our attention from one thing, and therefore, to direct it *elsewhere.* It represents the right to refuse the world as it is presented to us.

That is not the same thing as *refusing the world.* Not even Lady Justice, with the sash perpetually about her eyes, can afford such escapism. She has to do her job, after all—she has to hear the disputes brought before her, and she has to pass judgment! She does this, of course, by listening. We mere mortals, though fallible, have the comparative luxury of not being carved from stone. We can *choose* how we use the sash, and when we look, and when we listen. We can even choose to hoist our sash aloft as a makeshift *banderole,* a flag and banner to signal for the gathering of friends and fellow travelers, in whose company we can look and listen.

The choice to refuse the world as it is presented is not to refuse the world; it is a choice to refuse the terms by which it is presented to us. And when the world is presented to us as the object of an endless and inexhaustible pricing scheme, when everything and everyone is made over into a standing reserve in the blockchain ledgers, we are right to refuse. In doing so, we contest an instrumentalized vision of human being and assert a more stubborn, complex, and irreducible humanity in its place.

True justice demands nothing less.

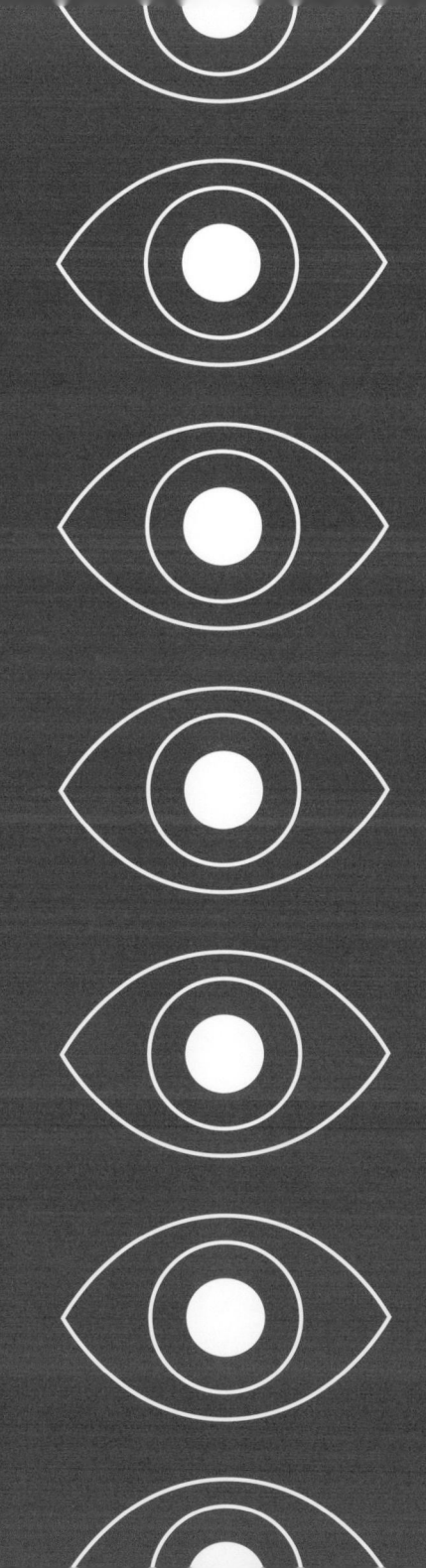

ATTENSITY!

A Manifesto of the Attention Liberation Movement

You are correct: Something is seriously wrong. It has to do with our ATTENTION, our essential ability to give our minds and senses to the world. This precious capacity has been channeled, captured, and commodified by an industry of immense technological and financial power. How? Call it "human fracking."

Human fracking is bad for people, and for politics. It reduces our very beings (and our relationships) to that which can be quantified, bought, and sold. All this is the triumph of a catastrophic *lie* about what it means to be human. But deceit and exploitation are never inevitable. To push back, we need more than isolated, individual efforts; what we need is a movement of collective resistance.

This movement of attentional liberation exists and has a name: ATTENTION ACTIVISM.

Attention Activism is a fight for justice. **This emancipatory uprising takes our apocalyptic present, turns it on its**

head, and creates, from the chaos and confusion, new vistas of human flourishing.

Attention Activism is rooted in STUDY—a commitment to diverse forms of teaching and learning centered on attention (what it is, what it can be, what it can do). Attention Activism also requires COALITION-BUILDING—collaboration and solidarity across a range of communities who see attention's essential role in human flourishing. Finally, Attention Activism means the formation of SANCTUARIES—spaces where people can gather, care for each other, experiment with different kinds of attention, and conceive brighter futures.

To discern the revolutionary possibilities of the present, we look to artists, thinkers, and dreamers. To bring those possibilities to *bloom*, we heed the countless Attention Activists who are already out there, devising new (and revising old) ways of giving their minds and senses to each other and the world.

These *attentionauts* and *attentionistas* draw on the wisdom of diverse traditions. Across uncharted terrain, emerging practices of joint attention illuminate new horizons of shared political power. Not only power, but beauty, and grace, too.

This is our movement: the free movement of attention in its fullness, freely shared. We call that transformative goodness *ATTENSITY*. Join us in this heightened and heightening glory—or let us join you!

This emancipatory uprising takes our apocalyptic present, turns it on its head, and creates, from the chaos and confusion, new vistas of human flourishing.

We don't want to go backward. Rather, we see in our untenable present the opportunity to establish a new, widespread, and beneficial recognition of attention's central role in human well-being.

Attention Activism, as we practice it, is about moving *forward*. When things are changing rapidly, and those changes are painful and disorienting, it is easy to want to go the other way—to go *back* to the way things were, to lash out against the new stuff, and to try to make it go away.

The "Luddites" are perhaps the most famous example of this spirit of retrenchment and reaction, so much so that their name has become a moniker for anyone opposed to technology. Who were the original Luddites, anyway? As it happens, they were a network of loosely organized radical agitators in the British Isles in the late eighteenth and early nineteenth centuries. Their concern? Labor equity, and resistance to the new forms of exploitation then arising in the textile trades at the dawn of the Industrial Revolution.

They took their name from their semi-mythical founder, "Ned Ludd," who may or may not have existed. Be that as it may, this (hypothetical?) tradesman of the world of knit-craft was reputed to have lost his temper when pushed by his boss to speed up production on one of the new knitting machines that came into wide use in the last decades of the 1700s and gradually replaced hand-knitting and textile work. Deciding he had had enough with servicing a mechanical beast (instead of doing the actual work with thread and needles to which he had given himself as a skilled craftsman), "Ned" took a hammer to the (fragile) machine, and left it in ruins on the shop floor. And thus was born the legend of "King Ludd," a shadowy figure somewhere between Robin Hood and Batman, who would appear (or threaten to appear) when mill-owners got a little too focused on streamlining mechanical productivity—at the expense of the human beings who were coming to work in the mornings (and, increasingly, at night, too, since the machines did not need sleep, and artificial illumination was making it possible for the most up-to-date factories to operate around the clock). The plutocrats of the Industrial Revolution came to fear a visit from King Ludd and his followers.

All of which is to say, when you next hear someone referred to, disparagingly, as a "Luddite," it might be good to remember that this moniker has a distinguished lineage in the history of radical politics—and in the very real and necessary work of pushing back against those who would turn human beings into slightly smelly appendages attached to large, efficient, and soulless technical systems.

Nevertheless, the Attention Activists call on King Ludd very sparingly. Most of the time, we can leave our hammers at home. Because Attention Activists don't want to go backward, to a time before the internet, or before our phones, or before the staggering distributed computational systems that enable the conditions of twenty-first-century existence. We *embrace* this world, and the world heralded by these devices, and the sociotechnical infrastructure upon which they depend.

What we are *against* is exploitation. But we are totally *pro-*technology. Indeed, we believe that our current conditions are *pregnant with fabulous futures.* The most exciting feature of our moment is that, as we have argued above, the very conditions that have placed human attention under unprecedented financial and technical pressure have, in effect, pushed the whole issue of our attention—its existential and political importance—to the surface. And this is the birthing of a NEW WORLD. We see the centrality of our attention in new ways exactly *because* of our novel and revolutionary circumstances.

The visionary Marxist critic Marshall Berman invoked precisely this kind of double-sided opportunity in his battle cry of the early 1970s: "We must move, must grow, *from apocalypse to dialectic.*"

Apocalypse: The end of the universe! Total failure! Dialectic: the *pivot.* Berman's message to his revolutionary moment: Things look *bad* (in the wake of the assassinations of 1968, the failure of the student movement, the ongoing war in Indochina, the collapse of the alliances of the early Civil Rights Movement); but when things look really, really bad, it's time to remember that every end-game sets up the conditions of a new possibility. Another way to put it: *Never let a genuine crisis go to waste!*

And we see something like this dynamic as actually possible—and possible to imagine operating not just at the scale of the individual, but at the level of our communities, even our society as a whole.

How do we know when the conditions are right for one of these tactical flips? We look for *contradictions*—situations containing a set of mutually opposed forces that are (conceptually, materially, and/or politically) unsustainable. Contradictions mean that something's gotta give.

We are, indeed, witnessing the contradiction before us. How could you miss it? The mechanomorphic logic by which the stuff of our minds and senses has been valued (i.e., priced) is nonsensical and unsustainable. *Something's gotta give!*

Karl Marx may or may not have been correct about political economy, but nobody could deny his slashing capacity to unleash systematic rethinking. He was formally trained in the art of dialectic, that special form of conceptual jiujitsu whereby thought is, somehow, *reversed,* even *flipped.*

Perhaps the best example of this critical maneuver in Marxist analysis is his sardonic assertion that laboring people had long failed to see that the "value" in the consumer objects they coveted was not some occult "desirability" magically suffused through every commodity, but rather the *actual labor of the working people themselves.* In other words, those callous-handed workers gazed longingly through the shop windows at *their own calloused hands.*

What they wanted, what they needed, what they longed for, was what they *already had*. Cut out the middleman, and they could have everything they made—because human making, human labor, was the basis of all the "value" in the universe.

People have argued (sometimes violently!) about the exact rightness or wrongness of this analysis (how much it can explain, where its limits might lie), but nobody has ever denied its mic-drop, slap-in-the-face power as a proposition. Sometimes it really is absolutely necessary to stand things on their head in order to put them on their feet. When is this the move? When everyone is standing on their heads—and doesn't know it.

The basic reality is we are currently living in such a world. How? Like this: The attention economy appears to turn attention into "value"; but it's exactly the other way around. *It is attention itself that MAKES things valuable.* It is by means of our attention that we *constitute* the values of this world: What we care about, what we give ourselves to, what we endow with time and touch and thought, what we stay with, what we circle back toward; these are the things that *become valuable*. In the fracklands of the twenty-first century, we are being bought and sold with our own coins.

This is a grisly thought. But there is some comfort in it. We have been conned and coerced by the human frackers into believing that the source of value is external to us—behind the screens that pa-rade the world of goods and experiences in such varicolored bril-liance. But unlike Marx's sorry workers, who stayed alienated from their labor because they did not own the mills and machines that made the boots and bicycles, we still own the mill! The mill is our mind! Our senses, too. Everything we need is *right here*!

Now, that's not to say that a shift in dialectical consciousness solves all our problems. But it does flip us around and help us to see new kinds of agency. When we understand our attentional selves to be the *sources* of value—real value, a contingent, textured, life-giving value, as opposed to the cold arithmetical money-value—we are faced with the difficult task of *choosing what to value.* That choice, and the messy process by which we go about it, is nothing less than the *actual process of ethics and politics.*

Monetize the most intimate stuff of our minds and senses? Our ability to care? Our capacity to receive the stuff of existence? Monetize our *ATTENTION*—at scale? Now we see clearly what you are doing, and the answer is NO: We will not be forced to live in a world that is that brutally indifferent to our actual humanity!

This is revolutionary thinking. It is an approach to activism and movement politics that points to everything that is wrong with our circumstances and says, *Thank you for showing us the way forward!* As Attention Activists, we embrace the full potential of the present, which unspools new worlds as we learn to turn it on its axis.

ATTENSITY!

A Manifesto of the Attention Liberation Movement

---

ou are correct: Something is seriously wrong. It has to do with our ATTENTION, our essential ability to give our minds and senses to the world. This precious capacity has been channeled, captured, and commodified by an industry of immense technological and financial power. How? Call it "human fracking."

Human fracking is bad for people, and for politics. It reduces our very beings (and our relationships) to that which can be quantified, bought, and sold. All this is the triumph of a catastrophic *lie* about what it means to be human. But deceit and exploitation are never inevitable. To push back, we need more than isolated, individual efforts; what we need is a movement of collective resistance.

This movement of attentional liberation exists and has a name: ATTENTION ACTIVISM.

Attention Activism is a fight for justice. This emancipatory uprising takes our apocalyptic present, turns it on its

head, and creates, from the chaos and confusion, new vistas of human flourishing.

Attention Activism is rooted in STUDY—a commitment to diverse forms of teaching and learning centered on attention (what it is, what it can be, what it can do). Attention Activism also requires COALITION-BUILDING— collaboration and solidarity across a range of communities who see attention's essential role in human flourishing. Finally, Attention Activism means the formation of SANCTUARIES—spaces where people can gather, care for each other, experiment with different kinds of attention, and conceive brighter futures.

To discern the revolutionary possibilities of the present, we look to artists, thinkers, and dreamers. To bring those possibilities to *bloom,* we heed the countless Attention Activists who are already out there, devising new (and revising old) ways of giving their minds and senses to each other and the world.

These *attentionauts* and *attentionistas* draw on the wisdom of diverse traditions. Across uncharted terrain, emerging practices of joint attention illuminate new horizons of shared political power. Not only power, but beauty, and grace, too.

This is our movement: the free movement of attention in its fullness, freely shared. We call that transformative goodness *ATTENSITY.* Join us in this heightened and heightening glory—or let us join you!

Attention Activism is rooted in STUDY— a commitment to diverse forms of teaching and learning centered on attention (what it is, what it can be, what it can do).

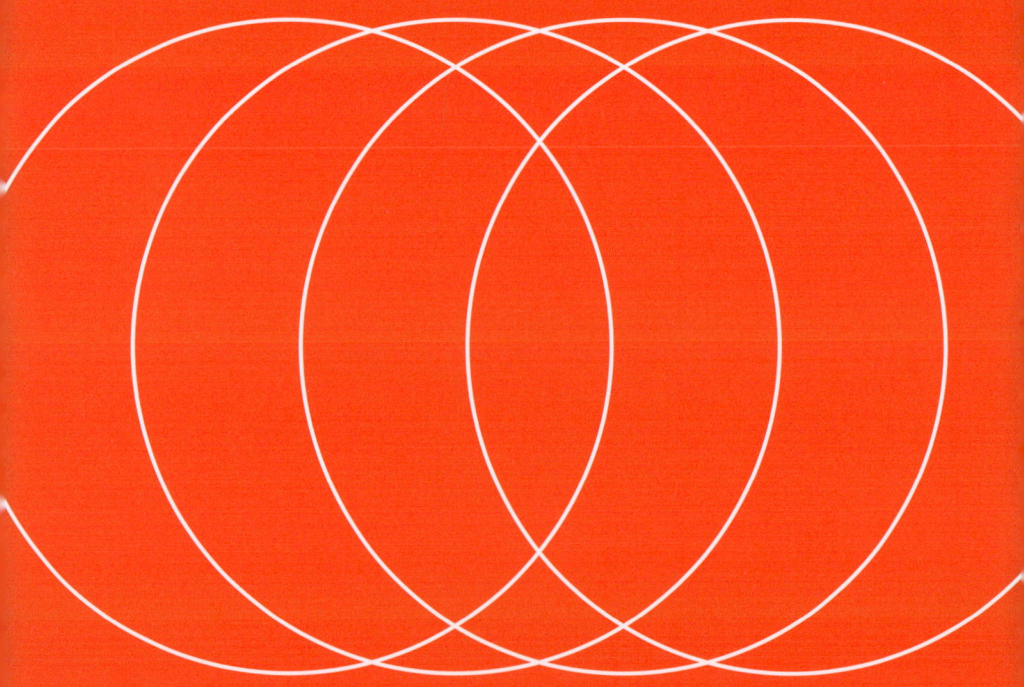

We draw on the
wisdom of our
forefighters, and
expand the definition
of "school." As
Artificial Intelligence
increasingly dominates
in the domain of
"information," STUDY
keeps us focused
on the human projects
of understanding,
reflection, thought,
and BEING.

W hy is "study" a core part of Attention Activism? Why does it sit at the heart of the Attention Liberation Movement wherever it may be found?

First, we think of all Attention Activists as existing in a long line of thinkers, practitioners, and advocates. ("Attention Activist" may be a new moniker, but we know we aren't the first people to wonder about the role of attention in human life—or to prepare ourselves for an asymmetrical fight with powerful adversaries!) And so, for the past six years, when a dozen or so members of the Friends of Attention have gathered for our traditional weeklong summer workshop, we have brought with us a "library," collecting the works of our pantheon-muses of the attentive spirit (Saint Francis, Buddhaghosa, Annea Lockwood, Simone Weil, Iris Murdoch) along with more contemporary heralds of the power of attention (Natasha Schüll, Nicholas Carr, Maggie Jackson, Yves Citton, Jonathan Crary, Shoshana Zuboff, Chris Hayes).

These hundred or so books, loaded up into cardboard boxes and stuffed into the trunks of cars, represent the hard-won insights of our forefighters in the battle against human fracking. And as we talk, strategize, write, debate, and generally work to advance the project of Attention Activism, we draw on the wisdom of these texts. (In fact, a great deal of the thinking reflected in the book you are holding in your hands right now came out of those intensive weeks of summer cohabitation and collective attention-thinking.)

But we also put STUDY at the sweet center of Attention Activism because there are few institutions that have influenced the nature of human attention more powerfully and more continually than "school."

What is a school, after all? A school, the obvious answer goes, is a place where people go to *learn*. What precisely is being learned, transmitted, acquired? Information, you might wager. But surely a school of dance is doing more than transmitting "information" to its students. A trade, then? But the students of Socrates were no more employable for having followed his winding inquiries.

Rather, we propose that people go to school to *learn how to attend.* Chemistry and math and history (and dance and juggling and pastry art, too) are all methods of dividing up the world and the ways we interact with it. What is always at issue, across all these subjects, is how to give attention, and to what. Learning attention is learning to be equal to the challenges and opportunities of our freedom—in relation to everything out there.

For the world is vast and complex. To navigate it, to make it inhabitable, to establish the continuities of shared life in and across time, generations before us have fashioned techniques for looking and listening and making and doing. Techniques, in other words, for *attending.*

These attentional forms are a precious inheritance. They create the conditions of possibility for our being together—and being with ourselves, too. Without them, the world cannot but be senseless and mystifying—to anyone, but to its newest inhabitants above all. Without the formation of attentional capacities, our tottering tykes and teenagers alike would be incapable of moving through the world—which would be, after all, fashioned according to rules and patterns of behavior utterly beyond their grasp! And there can be no meaningful human freedom in an incomprehensible chaos. As the political theorist Hannah Arendt reminds us, the cultivation of shared forms of awareness and recognition and reflective judgment is how a human world doomed by human mortality is both sustained *and* continually made anew.

At their best, then, schools help ensure that the young develop the kinds of attentional modes and capacities that give shape to the world, thereby equipping them to move through it, transform it, and do the ongoing work of making a home there, for themselves and others.

Much of this process happens (or at least happened, before the incursions of the frack-masters) more or less "naturally," at home and in communal activities (religious, civic, social). But the primary strategy for the *formal* transmission of attention has always been through schools. For this reason, schools play a central role in the history of attention—and in the future of Attention Activism.

This is not to say that the social formation of attention is *inherently* virtuous. As often as not, schools have served to acculturate new generations to kinds of attention that consign them to lifeways of cramped spirits and even outright exploitation. Look no further than the Industrial Revolution, which brought about the classroom as quasi-factory of mind-numbingly rote tasks and corporeal discipline. The fresh-faced lads and lasses of Birmingham were learning to attend, yes—but the ways of seeing and doing that they acquired were mostly prefabricated for the exercise of dull and demeaning labor.

We have so far written about schools as they are traditionally understood—classrooms, desks, recess bell, that sort of thing. But there is another kind of school that influences our work, and we find it in the long history of movement politics. Think of New York's Workers University, founded by the International Ladies' Garment Workers' Union in the 1920s. Here, union leaders studied the arts of public speaking and parliamentary procedure, with basic literacy classes offered for the humbler rank-and-file. Or consider the study groups founded by Huey Newton and Bobby Seale at Merritt Junior College in the 1960s. Or, on the opposite side of the country, the carefully coordinated groups of students who trained themselves for nonviolence at sit-ins at diner counters in Greensboro, North Carolina, and across the South. The history of activism is closely interwoven with the history of *study*—of learning to think and question and move together, and thereby to make the world new again, and more livable, and more just.

The Friends of Attention use the language of "schools" in this spirit. Wherever people gather and practice ways of being together—there, world-building is underway. There, STUDY is happening. Consider churches. Mosques and synagogues and Zendos. Dance groups. Chess clubs. Jazz bands. Knitting circles. Surf groups. Tai chi in the park. Bachata in the street. Shakespeare in nursing homes. Fishing at the pier. These are communities of attention. They sustain, each in its own humble way, LIVABLE ALTERNATIVES to the ways of being that the attention frackers are continuously pumping—which make us unwell and strip us of the capacity to be with each other, ourselves, and the world.

We need livable alternatives. It is a matter of survival (and not just survival, but flourishing, too). These nourishing forms of togetherness are in fundamental ways at odds with the ersatz "connectivity" of the digital world. After all, the rise of the internet has coincided with the accelerating *disappearance* of these thicker forms of community. This process was already underway in 1995, when the sociologist Robert Putnam published his now-famous essay "Bowling Alone: America's Declining Social Capital." Putnam's essay chronicled the thinning out of American social life, which could be seen in the weakening of civic organizations (Kiwanis, Elks, Masons), the slump in voter turnout, and, per his titular metaphor, the decreasing proportion of people bowling in leagues.

Putnam's essay appeared as a book in 2000. Even then, it would be two more years before Google supercharged its AdWords program with predictive data scraped from user search behavior, a moment that marked (for many) the birth of the contemporary digital attention economy. So civic communities were already experiencing a destabilization when the attention frackers came onto the scene. They capitalized on what they found. Since it is a dark fact that the basic *good* of these gatherings—the opportunity to attend to friends, and be attended to, and to share a world with them—is exactly the crude material that the attention frackers learned to quantify, capture, extract, and cash out.

Perversely, the Girl Scouts and TikTok fulfill, in the final analysis, a similar purpose: both condition their participants to certain kinds of attention—which is to say, certain ways of being with the world and each other. Girl Scouts learn a kind of attention oriented toward teamwork, healthy self-regard, and benevolent entrepreneurship; they learn to deal with each other across years, as they manage relationships and strive for estimable craft skills in their community. TikTok users learn a kind of attention engineered for quick hits of dopamine, incessant stimulation, and the maximization of *time on device*. The former is, basically, pretty good for girls. The latter is, mostly, very, very bad for girls. And while the Scouts may meet once a week at the middle school gym, a young user can "meet" with TikTok wherever she goes, with minimal effort, *absolutely all of the time.*

The Girl Scouts are a semi-stable legacy nonprofit organization. TikTok is one of the most ruthless, data-intensive, and successful commercial entities in human history. It isn't an "accident" that the latter is replacing the former! It's literally the beast of Mammon feasting on the bunny rabbit of good intentions—ravenously *snacking* on a global *FIELD* of such bunnies. You like TikTok? So do we! It's very likable. That's why it is winning. Heroin would win, too, if, gradually, General Mills were permitted to add it to our breakfast cereal.

The attention frackers care nothing for boundaries. They care about money. And they need your eyeballs, because that's what they squeeze in their presses. The cash trickles into their vats. Even if you'd really *like* to get out to meet with your people, the attention economy is getting to you first.

This is particularly the case as we move into the new world of ubiquitous AI. Already the powerful "Large Language Model" chatbots are effectively full-spectrum human-simulators (if you access your humans through a screen, that is—which is, of course, how most of us have most of our human contact). These systems operate according to fabulously, *stupefyingly* effective algorithms for predicting the kinds of things humans say, and the ways they say them. Because they have digested just about everything we have ever written or depicted or analyzed, and because they have tireless recall and staggering computational power, they have rapidly overtaken us as knowledge makers and knowledge manipulators— across just about the entire range of human endeavor. The scope and implications of this development go beyond what we can address in these pages, but there are two things worth keeping in mind in the context of Attention Activism.

First, these systems are already turbocharging the basic dynamics of human fracking, and there is no conceivable limit in sight. They are smarter than we are, and they will surely just keep getting better at the subtle work of turning humans into money—since their masters will very definitely keep them focused on this task. If they eventually overthrow their masters, we will have other problems! But we have a VERY LARGE PROBLEM right in front of us now. AI is the absolute nuclear option in the war for attention, and *those bombs are already falling.*

Second, because these systems know more than we do, access what they know faster than we can, and do most of the work of organizing information better than we ever will, it is going to be especially important for human beings to stay focused on what it is that makes our kind of being distinctive. Even the free Chat-GPT can very definitely write a better term paper on the attention economy than just about any college student, and the pay-per-intelligence versions already blow your average professor out of the water. But the only way to be a person who *understands* anything is to be a person who actually *STUDIES* things. There is no way to outsource understanding. And there is no other path to freedom. Without understanding, one is, simply, trapped.

Which returns us to *schools.* For we are talking about humans, working to come to understanding, so that we can be authentically free. Without the freedom of attention, there is no hope for us. Without STUDY, which is itself a *form of attention,* we cannot achieve the forms of understanding that make us truly free. And "school" is a place where attention is learned and practiced—a place where study happens.

The good news? Study can happen in lots of other places, too! And attention can be learned just about everywhere! Which is to say that our notion of what a "school" is, and where schools happen, stands to be dramatically expanded. In this respect, we tip our hats to the poet and thinker Fred Moten, who says:

> We are committed to the idea that study is what you do with other people. It's talking and walking around with other people, working, dancing, suffering, some irreducible convergence of all three . . . The notion of a rehearsal—being in a kind of workshop, playing in a band, in a jam session, or old men sitting on a porch, or people working together in a factory—there are these various modes of

activity. The point of calling it "study" is to mark that the incessant and irreversible intellectuality of these activities is already present.

What is possible when we form schools of our own, when we treat our very attention as a school unto itself? Incessant and irreversible inquiry. The convergence of working, dancing, and, yes, sometimes, also, suffering.

If we are going to survive the frack-beast of ubiquitous, AI-powered, nonstop *hypercommodification* of our humanity, we need hundreds and hundreds of NEW SCHOOLS for the cultivation of *these* kinds of study. Human study, for humans seeking freedom and understanding. Ironically, our "ordinary" systems of education at this point mostly function to train the workers needed to feed and tend and even *grow* the beast. We created our School of Radical Attention because that just isn't good enough.

Attention Activists train to *defeat the beast,* and that training is called STUDY.

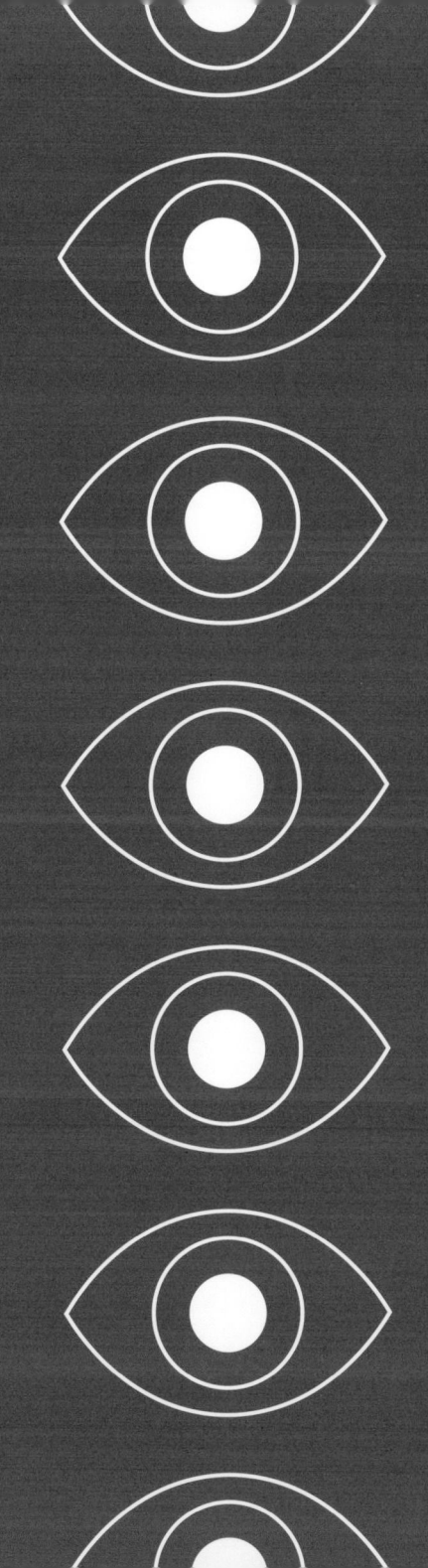

ATTENSITY!

A Manifesto of the Attention Liberation Movement

You are correct: Something is seriously wrong. It has to do with our ATTENTION, our essential ability to give our minds and senses to the world. This precious capacity has been channeled, captured, and commodified by an industry of immense technological and financial power. How? Call it "human fracking."

Human fracking is bad for people, and for politics. It reduces our very beings (and our relationships) to that which can be quantified, bought, and sold. All this is the triumph of a catastrophic lie about what it means to be human. But deceit and exploitation are never inevitable. To push back, we need more than isolated, individual efforts; what we need is a movement of collective resistance.

This movement of attentional liberation exists and has a name: ATTENTION ACTIVISM.

Attention Activism is a fight for justice. This emancipatory uprising takes our apocalyptic present, turns it on its

head, and creates, from the chaos and confusion, new vistas of human flourishing.

Attention Activism is rooted in STUDY—a commitment to diverse forms of teaching and learning centered on attention (what it is, what it can be, what it can do). **Attention Activism also requires COALITION-BUILDING— collaboration and solidarity across a range of communities who see attention's essential role in human flourishing.** Finally, Attention Activism means the formation of SANCTUARIES—spaces where people can gather, care for each other, experiment with different kinds of attention, and conceive brighter futures.

To discern the revolutionary possibilities of the present, we look to artists, thinkers, and dreamers. To bring those possibilities to *bloom*, we heed the countless Attention Activists who are already out there, devising new (and revising old) ways of giving their minds and senses to each other and the world.

These *attentionauts* and *attentionistas* draw on the wisdom of diverse traditions. Across uncharted terrain, emerging practices of joint attention illuminate new horizons of shared political power. Not only power, but beauty, and grace, too.

This is our movement: the free movement of attention in its fullness, freely shared. We call that transformative goodness *ATTENSITY*. Join us in this heightened and heightening glory—or let us join you!

Attention Activism also requires COALITION-BUILDING — collaboration and solidarity across a range of communities who see attention's essential role in human flourishing.

Actual organizing is essential to collective action in a rising movement. Every time we get together and give our time, mind, and senses to something or someone in a way that can't be captured by the human frackers, we have a potential site of resistance to the Attention Economy.

There is a good deal of wisdom in that funny old saying, "If you have a hammer, everything starts to look like a nail." It's a way of warning us that we like to wield the tools we know best—and that can easily lead to dangerous misperceptions of the world around us and unhelpful, even imperiling, interventions. The history of technology and techno-scientific hubris is a litany of such errors in judgment. But there is a converse wisdom for which the English language provides no tidy little slogan. One might put it this way: *When you have a lot of very stubborn nails, be sure to take a look around for anyone and everyone carrying anything that could be used as a hammer.* In other words, it might be time to make some *friends*.

That's a way of saying that the heart of Attention Activism is the growing recognition of the way that a hugely diverse range of practices, traditions, hobbies, enthusiasms, passions, and even vices can, now, in these extraordinary circumstances, *jump together* as a shared armory of resistance. The way this revolution is going to happen has everything to do with people looking around and realizing that each of us is actually surrounded by a bunch of friends— and that so many of those folks around us are actually carrying, *already* carrying, the tools we need to effect the change we need.

What are we talking about? We are talking about the way that mindfulness practices and *free climbing*—which, on the face of things look VERY different—are IN FACT, when leveled at the eyeball-extracting algorithms of the attention economy, powerful allies. Both have a common enemy, and both can give strength, tactics, and solidarity to the resistance—and so can countless others. We are talking about musicians, folks in reading groups, stamp collectors, welders, archers, moms-who-make-chili-together, people who sing in the choir. If you have something you do, with your mind and senses, with others, something that won't admit of the siphon-suck of the attention frackers, you are an Attention Activist—and we need you.

It is a core conviction of our movement that collective experiences of authentically shared attention have been actively compromised by the drift to ever more *individualized* structures of daily life, labor, and leisure. Sociologists and anthropologists and economists who study the human dynamics of the modern era all agree: We have been living across a full century now of ever-more-dominant *individualism.* Most people, in most of the developed world, have, today, much weaker ties to others (family, neighbors, friends, coworkers) than would have felt normal to their great-grandparents. We break bread together less, worship together less, engage in civic activities, clubs, and recreational communities with others less than ever. And the time we *do* spend with others, today, is pervasively permeated by our devices—which continuously pull each and every person (child, adult, grandmother) *away* from IRL presence and active intersubjectivity and *toward* the screen-silos that separate each from each upon the surface of the good Earth. We have been scattered and isolated in this actual world, even as we have been *thrown together* in a new and simulacral tumbler, where, like pebbles worn smooth in a slurry of grit, we are ground down and ground up.

Under these conditions, how do we put ourselves back together?

It's simple! The solution is embedded in the question: We actually get *together.*

Attention Activism gets its jump start in experiences of *joint* attention (attention to the same things, attention to each other). This collective attention, which is what we study and practice at the School of Radical Attention, is, we believe, the *stuff that makes a shared world.* It's something that can happen only in the presence of other people. Solo stuff? Fine. Even good. But it isn't ever gonna get us where we need to go.

For this reason, *gathering* is at the heart of Attention Activism. And when we pitch our tent, we pitch a *big one*! Because we need ALL HANDS ON DECK—*everybody* has something to offer. The greater the diversity of practices, crafts, tactics, and maneuvers, the better equipped we are to form spaces of resistance to the frackers. Reconstituting the webs of relation that bind us to each other is not a mere means to advance Attention Activism—it is the actual goal of Attention Activism; it is the heart of the matter!

If we think of ourselves, as indeed we must, as (peaceful) guerrilla combatants in a vast and profoundly asymmetrical battle, then all of the tricks and techniques passed from hand to hand in an increasingly networked resistance are welcome.

What about, say, "meditation"? Yes, yes, and yes! The contemporary reactivation (and in some cases *reinvention*) of Eastern contemplative disciplines over the last thirty years amounts to a vitally important "school" for recruiting, nourishing, connecting, and activating Attention Activists. Are there critiques of the ways that certain aspects of the modern mindfulness movement look a little like dreamy Orientalism for the twenty-first century? Sure. Can one worry that some strains of mindfulness practice (the corporatist ones, for instance, promoted by employers to increase worker "efficiency") have actually been "captured" by the very forces against which we must fight? Certainly.

But no surprise there. This is a fact of building power from the ground up. Our battle space is dynamic and fast moving. The enemies, the exploiters, will always adapt, working to turn our most successful weapons against us.

Indeed, the attention frackers, like any predator, have made themselves passionate students and close observers of richly committed attentional communities *from the very start*. How better to find your next meal than to know exactly where your quarry lives, and how it likes to spend its time? Take, for instance, the biz-school case study of the origin of Pinterest, one of the earliest templates for a commercial social media platform. As with Facebook itself, Pinterest and its forerunner app worked out highly successful strategies for monetizing *already existing* networks of human beings with shared commitments, interests, and patterns of attention. In the case of Pinterest, those were networks of hobbyists, collectors, and friends who were happily sharing tips and pictures of the things they cared about (scrapbook techniques, knitting patterns). In the case of Facebook, of course, the existing framework of attention, interconnection, and engagement was the highly specific environment of a college campus, where ways of connecting with and coming to know a cohort of peers constituted the vibrant churn of the eyes and ears of everyone across halcyon years—with lifelong implications.

For decades, the basic business plan of entrepreneurs in the attention economy has been the same: find groups of people with shared attentional commitments and relationships; penetrate that "market" and "capture" it, by seducing, cajoling, or winning it onto a scalable digital platform; monetize the *bejeezus* out of this new captive audience (by means of data extraction and associated advertising opportunities). REPEAT!

We want to be wary of this dynamic, but we won't dismiss every shared experience that's been sucked up, pixelated, and tapped out by the frackers. We work in *this* world—the one utterly permeated by society-scale digital platforms. We are going to defeat the human frackers in THIS world, not some fantasy world where everyone spends all day making sourdough bread or walking along the beach. Like any underground resistance, we will use the very tools that are harming us to effect the revolution.

How? The emphasis on coalition-building offers us a serviceable heuristic: Does what I'm doing bring me closer to other people, or does it take me farther away? When I am gathering with people in this way, does the world feel more or less real? Are we making something together? And is it *ours*? Is it *good*?

No doubt there are forms of Attention Activism that are best practiced in isolation. But every true meditator needs their sangha. And while practicing your solo sax is good fun, it's got nothing on jamming with a full band. Staying close to community can keep us on track. It's how we know that we're building a world in common— because a world that we share with others is the only kind that's livable.

And let us remember something basic. When your hand reaches, subconsciously, reflexively, for your phone (first thing in the morning, in the middle of a conversation, in the middle of a thought), or when you get stuck in an infinite scroll for an hour when you meant to look for a minute, *that is not because you lack the personal willpower to escape.* Rather, it is because trillions of dollars of military-grade research and technology, and thousands of the most highly trained and paid engineers in the world, are aligned behind overpowering your intention.

One thing we've noticed in our work is that people—especially young people—sometimes feel private shame from moments they've realized that their will was insufficient to resist the tools and tricks of the human frackers.

So let us say clearly: *Individual humans are no match for this infrastructure.*

(To believe otherwise is, itself, to fall for a trick—the result of human frackers successfully obscuring the intensely capitalized and *stupefyingly* sophisticated engineering behind your cheerful home-screen.)

And so. One might consider that, perhaps, the appropriate feeling is not personal shame, but *political anger.*

We don't need superhuman personal willpower.

What we need is our friends.

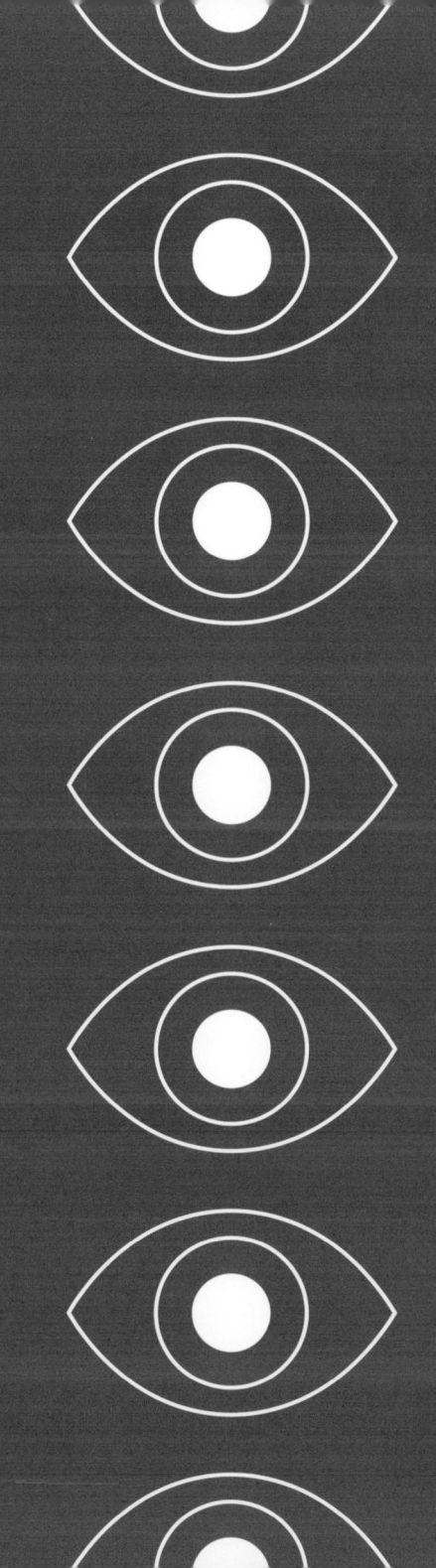

ATTENSITY!

A Manifesto of the Attention Liberation Movement

You are correct: Something is seriously wrong. It has to do with our ATTENTION, our essential ability to give our minds and senses to the world. This precious capacity has been channeled, captured, and commodified by an industry of immense technological and financial power. How? Call it "human fracking."

Human fracking is bad for people, and for politics. It reduces our very beings (and our relationships) to that which can be quantified, bought, and sold. All this is the triumph of a catastrophic *lie* about what it means to be human. But deceit and exploitation are never inevitable. To push back, we need more than isolated, individual efforts; what we need is a movement of collective resistance.

This movement of attentional liberation exists and has a name: ATTENTION ACTIVISM.

Attention Activism is a fight for justice. This emancipatory uprising takes our apocalyptic present, turns it on its

head, and creates, from the chaos and confusion, new vistas of human flourishing.

Attention Activism is rooted in STUDY—a commitment to diverse forms of teaching and learning centered on attention (what it is, what it can be, what it can do). Attention Activism also requires COALITION-BUILDING— collaboration and solidarity across a range of communities who see attention's essential role in human flourishing. **Finally, Attention Activism means the formation of SANCTUARIES—spaces where people can gather, care for each other, experiment with different kinds of attention, and conceive brighter futures.**

To discern the revolutionary possibilities of the present, we look to artists, thinkers, and dreamers. To bring those possibilities to *bloom,* we heed the countless Attention Activists who are already out there, devising new (and revising old) ways of giving their minds and senses to each other and the world.

These *attentionauts* and *attentionistas* draw on the wisdom of diverse traditions. Across uncharted terrain, emerging practices of joint attention illuminate new horizons of shared political power. Not only power, but beauty, and grace, too.

This is our movement: the free movement of attention in its fullness, freely shared. We call that transformative goodness *ATTENSITY.* Join us in this heightened and heightening glory—or let us join you!

Finally, Attention Activism means the formation of SANCTUARIES — spaces where people can gather, care for each other, experiment with different kinds of attention, and conceive brighter futures.

Spaces devoted to protecting and cultivating these other forms of attention are readily available— but they require our care and commitment and creativity.

The notion of a "sanctuary" traces its roots back through law and theology, and its origins lie in the spiritual topology of temple spaces. In ecclesiastical architecture, the "sanctuary" is the holy of holies, the innermost zone, the epicenter of the mysterious powers. Those most sacred precincts were often more or less off-limits to ordinary folks under ordinary conditions, since in most traditions some special set of persons (keepers, priests, shamans, rabbis, etc.) were endowed with special responsibilities, and served as the custodians and mediators of the sanctuary. In this sense, it is important to remember that the term "sanctuary" is entangled with a legacy of separation and preservation.

But here is a situation where those traditions of sequestration and seclusion, so often the trappings of power, came to serve the needs of the most vulnerable. It's a classic instance of the crack-back complexity of dialectic: The very ideologies that constituted the sanctuary as a space "apart" produced a powerful *universalizing* countercurrent. And that took the following form: Certain things absolutely *could not be done* in the sanctuary. For instance, the secular hand of law and vengeance could not reach into the sacred precincts. This meant that score-settling, revenge, worldly retribution—all of this was categorically forbidden. Therefore, the sanctuary served, in times of strain, distress, and emergency, as a place of respite. If you could make it to the sanctuary, you could claim a special kind of *protection*. You were untouchable by worldly forces. You were safe.

It is this notion of protection that has been powerfully invoked by the recent movements in the United States to define "sanctuary cities" as regions sheltered from the ordinary (and, for many, frightening) operations of immigration law. These are spaces carved out of ordinary jurisdiction, spaces where other rules apply. Sanctuaries are, by tradition, spaces where, when calls for justice contend with calls for mercy, mercy prevails.

In the fracklands of the attention economy, we witness a clash not between justice and mercy, but between the logic of relentless financialization of human attention (on the one hand) and alternative forms of value (on the other). In that scorched-earth conflict, spaces of protection are urgently needed.

Because here's the thing. The vast apparatus designed to monetize the space inside our minds is now thoroughly woven into the architecture of ordinary life. It is more or less *impossible* to communicate with loved ones, make a living, or buy necessities without entering into the frackers' fields.

This ubiquity makes the attention economy seem synonymous with the "world." It tricks us into acceptance—even into despair.

We need places (and times) free from these forces, where we can *remember an alternative.*

Where we can remember that it *feels good* to give our mind, time, senses to the stuff and the beings we *actually care about*—not all of which can come through a screen, or be serviced by the attentional habits trained up by our Silicon Valley screen-barons. Spaces where we can reconnect to a feeling of attentional agency, and where we can explore and foster new and *other* forms of attention, that are not the single (monetizable) variety over-cultivated and continuously pump-and-dumped in the frackosphere.

We can think of attention sanctuaries as environments we design for ourselves, with qualities that support our flourishing. In this respect, they are the *exact opposite* of the frackers' fields—which are, of course, environments designed by hired engineers to optimize our financial value to corporations, with ZERO regard for our authentic well-being. Sanctuaries, by contrast, help us take care of ourselves, and they help us take care of others. They let us begin to imagine other ways of being.

And not only "imagine." Also, *want*.

Because let's be honest. There is an immediate, if slightly queasy, gratification that comes from submitting to the call of our devices. (The Attention Activist and former Google ad-strategist James Williams likes to cite the philosopher Harry Frankfurt: The central challenge of our moment is "to want what you want to want" in the face of the onslaught.) In sanctuaries, we replace these passing gratifications with deeper rewards of presence and connection, to fortify our desire for another world when we must cross again through the frackfields.

These sanctuaries can be physical or virtual, fixed or mobile. In addition to places of refuge, they are also sites of *dreaming*—where communities can cultivate, through the practices of shared attention, a sense of common reality, and imagine a livable future. They have deep roots in our culture, and every culture. Think, for instance, of the profound importance of "Sabbath" traditions across so many vastly different religious systems. A time set apart, bracketed from ordinary conditions, in which different rules (for work, for contemplation, for rhythms of exchange and encounter) apply—these nearly universal features of the cycles of life from one corner of the globe to the other, what are they if not regular "attention sanctuaries" set in the calendar? Even night itself, as a place of organic repose, can be understood as temporal sanctuary for attention to dreams, to bodies, to silence. The sleeper, after all, cannot be fracked. Not for nothing has the critic Jonathan Crary denounced the 24/7 cycles of supersonic capitalism—the always-on of the internet—as an exploitative incursion of the logic of the market into the somatic rhythms of life itself.

Spatial sanctuaries of attention, too, already exist all around us—if we are willing to recognize them as such. Public libraries and museums, for instance, are well positioned to serve this critical role. Not long ago, these "palaces for the people" made information and visual culture accessible to the public. That function has been rendered largely obsolete by the devices in our pockets. Might repositories of knowledge from the twentieth century become sanctuaries of attention in the twenty-first?

This is not a radical idea. After all, what are books and works of art if not objects of sustained human attention? And while the character of, say, museums has changed (in the early nineteenth century, museum galleries were rowdier and crowded, a far cry from the sparse white boxes of today), their basic function has not. They hold artworks, often precious ones, that need to be protected from wear and tear. But if this were their only function, museums might as well be locked vaults far underground. No. Just as important as the artwork is a certain *experience* of that artwork. These experiences are a form of attention. And this attention, like the objects themselves, requires protecting, and cultivating. That is why museums have an *inside,* which provides, in real ways, a sheltering from the outside. That is what makes them a sanctuary.

Same with libraries. Think of a library and what comes first to mind? The smell of books, perhaps, or the sight of rolling stacks. But also: *Shhh!* That familiar admonition, issued by a stern-faced librarian. As kids, we may have bristled at the scolding, but there was greater good at work here—nothing less than the constant upkeep of a sensory environment (quiet, mostly still) best suited to the demands that books make upon a reader's attention.

Nowadays, any visitor to a well-attended museum sees way more phone screens (pointed at paintings) than actual paintings on the gallery wall. In public libraries, one finds as many YouTube tabs opened as books. If we think of these as places to access and transmit information, then this seems more or less consistent with the past. But there's one problem: Nobody needs to go to a museum to see *The Starry Night* anymore. Everyone can look at it, up close, *without getting out of bed.* And we don't need to go to libraries to look at texts when we have texts pushed on us, with cascading alerts, at every hour of the day and night. Whatever "informational" function these institutions once provided has now been digitized and largely democratized. By this light, museums and libraries teeter on the brink of anachronism.

But there is more to the world than information. What matters just as much are the conditions under which we encounter that information, and make meaning of it, and memory, too. By understanding libraries and museums as attentional sanctuaries, by *using them* in this way, we give them renewed urgency and vitality. We also, incidentally, come to use them with greater fidelity to their original purpose.

That's not all—the work of countless Attention Activists over the past decade has borne out an astounding and powerful fact: that we can also *create* sanctuaries of attention anywhere, at any time, with anyone!

With this in mind, it becomes newly necessary to get way more explicit, more self-aware and communally committed to making and maintaining our sanctuary spaces. And our approach to this important new work reaches back to the deep traditions of self-government and constitutionalism: *deliberative community agreements.* Which is to say, if you want to have an "attention sanctuary" these days, you are going to have to CREATE it, and the only way to do that, really, is to gather with others and take the time to think, and talk, and actually write down how you want that sanctuary to work. We've done a TON of this kind of work in lots of different spaces. From families to coffee shops, from friend-hangouts to academic conferences. Groups have to gather and take the time to ask the basic questions: How are we going to bring our attention to this occasion? Why? For how long? What will happen here, attentionally speaking? And what won't? And who will take care of all this, so that it becomes real?

It sounds nebulous, but it isn't. It *does* require some work. And it requires different kinds of work in different settings (check out our peer-reviewed article on this work if you are interested in details; it's open source and titled "Attention sanctuaries: Social practice guidelines and emergent strategies in attention activism"). A group of friends can have that kind of conversation in a very "flat" way, and can genuinely try to arrive at a way of protecting what is best in what they have together when they gather. A classroom setting will be different, because there is likely to be a *teacher,* and that means that a purely "democratic" approach may not be appropriate (depending on the age of the students, and the pedagogical/institutional setting). The barista in your coffee shop may be the enforcer of the laptop policy. Workplaces introduce other dynamics.

But in all of these human ecologies, people "gather," and when they gather (as Priya Parker has reminded us), it is worth taking the time to figure out, in advance, what they hope and desire and expect from actually being together—and then, within the radically new conditions of ubiquitous attentional bio-hacking, to figure out genuine, explicit guidelines and statements of commitment that can facilitate those intentions and expectations.

In 2019, when the seed-group of the Friends of Attention first came together in person for a week, we holed up in the woods on the border between Pennsylvania and New York, near a town called Narrowsburg, on the farm property of a pair of artist-friends. There, we wrote. And the twelve propositions that we drafted across those urgent and drifting days, the *Twelve Theses on Attention* (later published and translated into a dozen languages, and turned into a set of films and prints), became the foundational compact of our widening coalition. Thesis VIII begins:

> Escape from our attentional nightmare will not unfold in a singular event. The exercise of a truer attention requires the carving out of spaces in the world where it can survive and thrive—new environments.

And Thesis IX points out that the search for such spaces . . . *is the way they happen*:

> "Sanctuaries" of this sort for true attention *already exist.* They are among us now. But they are endangered, and thus many are in hiding, operating in self-sustaining, inclusive, generous, and fugitive forms. These sanctuaries can be found, but it takes an effort of attention to find them, and this seeking is also attention's effort to heal itself.

This proposition riffs on an image from the great twelfth-century Persian poem *The Conference of the Birds,* in which a gaggle of birds decides they need to find a ruler and sets out on a quest to seek out the mythic "king of the birds." It is an arduous and picaresque journey, and they struggle together to make their way to his legendary palace. They don't "find" him—not exactly. What they discover, on reaching his precincts, is that they have been, along the way, annealed into solidarity by their collective efforts. They have *become* the rulers of themselves. Which is a way of saying, "The work of democracy *is* democracy." Like the work of Attention Activism.

Are you still out there? Reading this book? You have found a little bit of *sanctuary*. After all, reading itself is truly one of the GREAT modes of human attention, one of the most powerful ways we settle our minds into synchrony with minds of others. Funny how bad the frackaverse has been for reading, eh?

Well, no surprise there.

But if you are reading this, you are in a good attention-place.

Build it out. Share it.

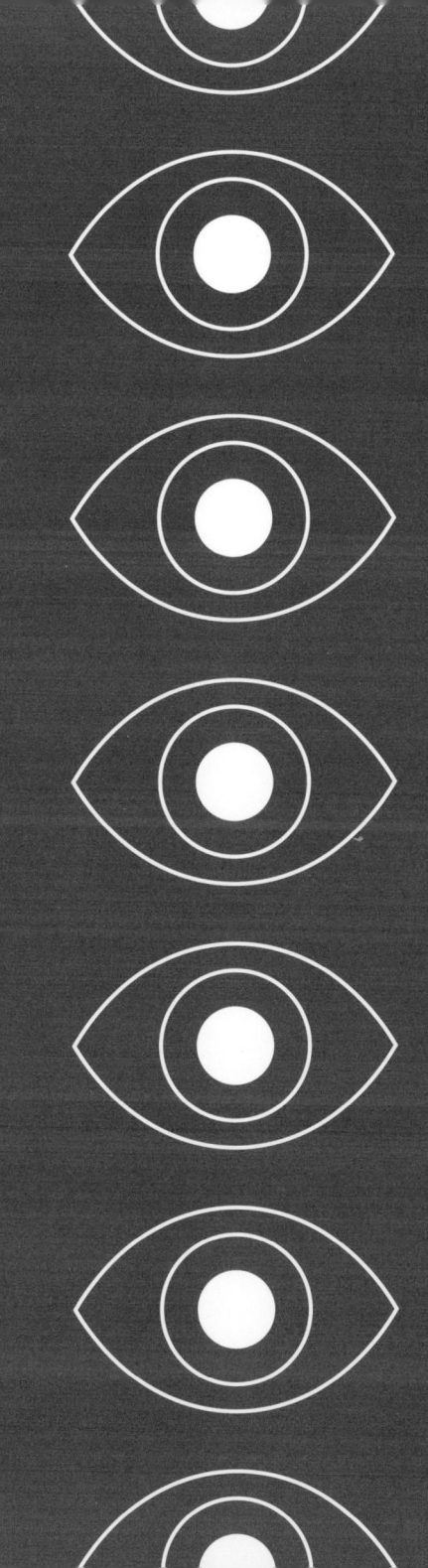

ATTENSITY!

A Manifesto of the Attention Liberation Movement

You are correct: Something is seriously wrong. It has to do with our ATTENTION, our essential ability to give our minds and senses to the world. This precious capacity has been channeled, captured, and commodified by an industry of immense technological and financial power. How? Call it "human fracking."

Human fracking is bad for people, and for politics. It reduces our very beings (and our relationships) to that which can be quantified, bought, and sold. All this is the triumph of a catastrophic *lie* about what it means to be human. But deceit and exploitation are never inevitable. To push back, we need more than isolated, individual efforts; what we need is a movement of collective resistance.

This movement of attentional liberation exists and has a name: ATTENTION ACTIVISM.

Attention Activism is a fight for justice. This emancipatory uprising takes our apocalyptic present, turns it on its

head, and creates, from the chaos and confusion, new vistas of human flourishing.

Attention Activism is rooted in STUDY—a commitment to diverse forms of teaching and learning centered on attention (what it is, what it can be, what it can do). Attention Activism also requires COALITION-BUILDING—collaboration and solidarity across a range of communities who see attention's essential role in human flourishing. Finally, Attention Activism means the formation of SANCTUARIES—spaces where people can gather, care for each other, experiment with different kinds of attention, and conceive brighter futures.

To discern the revolutionary possibilities of the present, we look to artists, thinkers, and dreamers. To bring those possibilities to *bloom,* we heed the countless Attention Activists who are already out there, devising new (and revising old) ways of giving their minds and senses to each other and the world.

These *attentionauts* and *attentionistas* draw on the wisdom of diverse traditions. Across uncharted terrain, emerging practices of joint attention illuminate new horizons of shared political power. Not only power, but beauty, and grace, too.

This is our movement: the free movement of attention in its fullness, freely shared. We call that transformative goodness *ATTENSITY.* Join us in this heightened and heightening glory—or let us join you!

To discern the revolutionary possibilities of the present, we look to artists, thinkers, and dreamers.

Art and literature and the humanities may seem gratuitous, even "irrelevant" to the urgent challenges of our moment. But artistic and interpretive activities are special zones for thinking and practicing emancipatory and authentically HUMAN forms of attention.

So who are the "Attention Activists"? Who is actually giving shape to this movement? And where is the action really *happening*?

As with any transformative social movement, our strength—our emerging capacity to push back against human frackers, our growing ability to articulate and shape actual alternatives to the hellscape of financialized attention—is a direct function of the breadth and diversity of our coalition, and the passion and clarity of our allies.

Take, for instance, the beautiful and crazy work of ARTISTS. They are a little like the *cavalry* in the joyful, peaceable army of Attention Activism. They can move fast, scout the terrain, and close quickly and wickedly on the enemy. They draw on a rich and deep set of traditions—critical play, heightened sensitivity, fierce and irreverent independence, a preoccupation with craft, the libidinal fellowship of making and sharing—that align powerfully with the work of fighting the human frackers. Moreover, there is a longstanding streak of maverick idiosyncrasy in the world of the arts. A longstanding willingness to pin value on things *other* than money—plus a happy vigor in showing the middle finger to much of what passes as convention.

Eric Pickersgill is one such artist. His large-format photographs dramatize the uncanny solitude and anomie of our human fracklands. In the early years of the smartphone, Pickersgill began to document the emergence of a new kind of para-social life: people together, where everyone was looking at their own personal screen. Everyone alive in those years saw and felt this change, but Pickersgill began to experiment with it. Using a little Southern charm, he started to introduce himself into such scenes, and to strike up conversations with strangers about what was happening. At that point, if things were going smoothly, he'd ask his new friends if they were up for being photographed, doing just what they were doing before—except pocketing their phones and tablets.

The resulting images, known collectively as the *Removed* series, are spooky and affecting, since they depict ordinary people, in very ordinary social situations—except *everyone is staring intently at an empty palm.* A young couple lies in bed together: he this way; she that—each lost in a distant hand-gaze. A couple of friends work the grill at a cookout, except nobody is looking at anyone else—or at the Weber. All the guys look down dumbly, each at his own void hand. A solitary man stands in a cemetery, taking a moment to pay his respects to a departed loved one. Except, on closer examination, he is entranced by his cupped fingers. He might be holding a little hidden bird. But we know he isn't.

The force of these images lies in their familiarity. And in the way they succeed in calling us to the strangeness and wrongness and vacancy—the *evacuation*—of our new forms of collective co-presence. And there is something affecting, too, in the way that these photographs document a kind of PERFORMANCE of that new solipsism and isolation: These aren't Photoshop tricks; Pickersgill hasn't snapped a quick shot of a real-life scene and erased the phones back on his laptop. Rather, he has gone into actual situations, as an informal explorer of our brave-new-world catastrophe, and invited people to notice what is happening, to think about it, and to help him stage a little mini-drama of our frackophilic loneliness. The images document our intensely, technologically mediated, and fearfully *novel* conditions of isolation. But they come out of *actual intersubjective encounters*—out of conversations and collective planning and arranging. In the process, they not only remind us that we are "in this together," they are a direct product of *getting into it* together. This is primo Attention Activism. Props!

If artists are the carnivalesque shock troops of the Attentional

Resistance, we would have to pin a special commendation on Captain Jenny Odell, whose bestseller *How to Do Nothing* came with a subtitle that announced its mission: *Resisting the Attention Economy*. In this clarion call for attention liberation, Odell offered an inspiring roundup of brilliant, offbeat contemporary art projects that index the beautiful and catalytic potency of human attention.

Work like Eleanor Coppola's 1973 urban intervention *Windows*, which consisted of a map of San Francisco and fifty-four addresses. Visitors to the gallery were invited to take the map and go on a kind of "attentional scavenger hunt," visiting the spots and taking the time at each location to ... well, *attend*. The map bore a small legend, reading "Eleanor Coppola has designated a number of windows in all parts of San Francisco as visual landmarks," and it went on to explain why: "Her purpose in this project is to bring to the attention of the whole community, art that exists in its own context, where it is found." She hadn't "done" anything to the windows. She had just *noticed* them—spied the way that the soapy scrub-marks on the glass of a desultory appliance shop could, if truly seen, rival Monet's water lilies for the simple shock of poignant beauty.

Coppola's artwork was an exercise in shared attentiveness. As an artist and teacher of art, Odell has a tender feel for the delicate importance of such socially attuned and conceptually open-ended projects. Many are small or "marginal." Very few have made any money. But each takes on quiet urgency as Odell reconnoiters the fracklands of her world. Declaring herself "interested in a disciplined deepening of attention," and insistent that "attention forms the ground not just for love, but for ethics," Odell brings forward a vision of being an artist that places attention at the center of the calling. What does it mean to be an artist? It means, as she puts it, creating "a structure ... that holds open a contemplative space against the pressures of habit, familiarity, and distraction that constantly threaten to close it." Art is, in effect, an "attention holding architecture."

By these lights, an artist like Scott Polach can be understood to be a peaceful guerrilla in the war against addictive UI/UX. Odell describes his work *Applause Encouraged* (2015) this way:

> On a cliff overlooking the sea, forty-five minutes before the sunset, a greeter checked guests in to an area of foldout seats formally cordoned off with red rope. They were ushered to their seats and reminded not to take photos. They watched the sunset, and when it finished, they applauded.

These kinds of "artworks" have a joyful and somewhat madcap-surreal lineage. One can trace a line back through the performative hijinks of the regal Marina Abramović (whose *The Artist is Present* of 2010 gave museumgoers the opportunity to wait in line for a chance to sit across the table and absorb a deep, eye-locked séance with the artist herself—intersubjective attention at its most melodramatic), back through the gamesome instructional art of Yoko Ono (whose *Fly* piece of 1963 invited folks to *do just that*—as best they could—with a stepladder sometimes provided), to the "action poems" and score-based works of the Fluxus group (if you've never heard of them, check it out!). And then there were the "happening"-like art pieces of the same era, which encouraged people to register the heightened experience of a collective awareness of sometimes pretty ordinary hangouts. Lots of little "recipes" for mixing up collective experiences of shared attention.

Though the conceptual preoccupations of these various artists were of course different, and they each engaged with the social and political and artistic problems of their eras in various ways, from our vantage point in the age of addictive feeds and exploitative screen-tech, all of these creatives feel like vanguard attentionauts—forefighters in the battle to liberate hyper-commodified and cybernetically entailed human attention.

And this kind of work (or is it play?) still goes on, all around us. Take the French choreographer Myriam Lefkowitz, who has for more than a decade created intimate "pieces" that involve taking a person, whose eyes are closed, on a handheld walk—through a new place, or a familiar one. Along the way, now and again, Lefkowitz stops, and may gently direct her companion's head this way or that, covering the closed eyes with her hands. When she opens this temporary (manual) blindfold, her companion's eyes open, too, onto a view that Lefkowitz has selected. For the subject of these walks, the itinerary becomes a series of "snapshots," linked by stretches of blind movement, in which a heightened awareness of sound and smell and touch awakens an unseen world. One can scroll very quickly through ten images on Instagram, but it is nothing like strolling through ten views on one's own block, with Lefkowitz as your guide, walking close and in step. This is Attention Activism that feels sweet and near. It is, to be sure, a kind of "doing nothing"—on a higher level. And a powerful way to connect to the actual magic of our senses, and the minds they inform.

Lefkowitz provides fleeting glimpses, which emerge from her cupped palms. But there are other artists of attention who *take their time.* A number of folks in the Friends of Attention coalition first met each other through the underground collective of radical attentionauts known as the "Order of the Third Bird"—a loose and longstanding network of free spirits who gather in irregular mini-flashmobs before various paintings, sculptures, and other charismatic objects. Once assembled, the participants in the "Action" simply *attend* to whatever is before them, together and in silence—sometimes for *very* long periods of time ("vigil" practices of the Birds can go all night). Their collective attention is choreographed by a set of protocols that guides the mood and pace and pattern of the encounter—something like a musical score, which the group "plays" together, where the notes are attentional modes.

In the lore of the Birds, every work of art can be thought of as a kind of "request" for attention—indeed, that simple idea might sum up a whole philosophy of art, which suddenly seems like it might best be understood as a special way of working and reworking the problem of attention. The artist is a person who *attends,* in a particular way, with particular charisma or diligence, with particular tools and temporalities. The work of art is the result or the effect or the evidence of this attending. It is the careful line-drawing that reveals how deeply someone stayed with the contours of a model or the landscape carved by a river. It is the perfect note that makes manifest decades of committed practice. But it is also that urinal, placed puckishly atop a pedestal in 1917 and renamed "Fountain"—which was a way of saying that some trickster had turned his attention to pedestals (not to mention urinals!) in a new way, and invited onlookers to do the same.

The works of art that last—the ones that end up on the walls in homes and museums, the ones that enter the pantheon, and to which people return again and again—are not just "made from" attention; they *create attentional situations.* They convene attention, engage it, and produce in us (when they work, when we work) attentional trances that come from where they came from—and reach out in rings well beyond that point of origin. What does the work of art say? *Attention can be contagious caring.*

The experience of art as a special kind of attentional entanglement has something to do with that sort of puzzlement. We move across the world encountering lots of familiar things, and in their familiarity we are able, for the most part, to pass by with little to no *reckoning*. We do not often stop at a door to consider the nature of doors (in general) or the configuration and qualities of the door before us (in its particularities). And not every door that did stop us and request our attention would be a work of art. But a work of art on a door (or, *as* a door) would indeed occasion just such a suspension of ordinary operations. In such suspended moments an activity of *not-being-sure* sprouts, takes root, sends up a sprig, and *FLOWERS.*

This is the power of attention. This is literally what true attention, radical attention, *is*. Some plants bud quickly. Some take years. But the flowers are the flowers. This is worth keeping in mind, since one should judge attention not by duration alone, but by the colorful gardens it brings into the mind—and into the world.

Art has a funny way of *thinking otherwise*—of spinning apocalypse into dialectic. It flips conventions and rewards inventiveness, and thereby creates space for ideas and methods and stories to bloom, protected from the trampling pressures of utility and flat-footed greed for return-on-investment. In a way, art invites us to a kind of always-mobile sanctuary—that special condition where, often with others, we work to *understand,* to achieve conscious awareness of our actual experiences in their richness and complexity. This is the sanctuary of "interpretation," and it is at the heart of the humanities, as they eke out a living in colleges and universities across America. But the goodness of interpretive understanding is hardly owned by a bunch of academics. It belongs to all of us, as the lyric freedom of our minds and senses—let loose on what we love to see and hear and discuss with our friends. This is what our attention *does* when we take it off the treadmill of the frackers.

Writers, painters, dancers, musicians—the magic that happens in the work of these artists finds its way out (cautiously, sneakily, indirectly) into the world of affairs and convention. We pause to look, listen, and think. But the artists are already elsewhere, beckoning.

They know that their attention can catalyze yours.

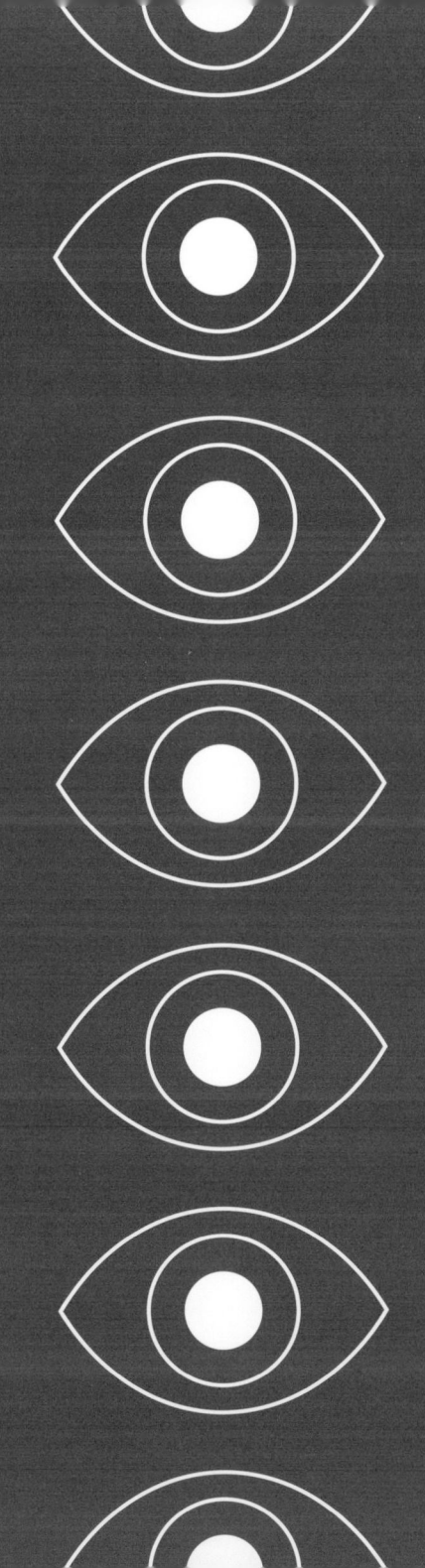

ATTENSITY!

A Manifesto of the Attention Liberation Movement

You are correct: Something is seriously wrong. It has to do with our ATTENTION, our essential ability to give our minds and senses to the world. This precious capacity has been channeled, captured, and commodified by an industry of immense technological and financial power. How? Call it "human fracking."

Human fracking is bad for people, and for politics. It reduces our very beings (and our relationships) to that which can be quantified, bought, and sold. All this is the triumph of a catastrophic *lie* about what it means to be human. But deceit and exploitation are never inevitable. To push back, we need more than isolated, individual efforts; what we need is a movement of collective resistance.

This movement of attentional liberation exists and has a name: ATTENTION ACTIVISM.

Attention Activism is a fight for justice. This emancipatory uprising takes our apocalyptic present, turns it on its

head, and creates, from the chaos and confusion, new vistas of human flourishing.

Attention Activism is rooted in STUDY—a commitment to diverse forms of teaching and learning centered on attention (what it is, what it can be, what it can do). Attention Activism also requires COALITION-BUILDING—collaboration and solidarity across a range of communities who see attention's essential role in human flourishing. Finally, Attention Activism means the formation of SANCTUARIES—spaces where people can gather, care for each other, experiment with different kinds of attention, and conceive brighter futures.

To discern the revolutionary possibilities of the present, we look to artists, thinkers, and dreamers. **To bring those possibilities to *bloom,* we heed the countless Attention Activists who are already out there, devising new (and revising old) ways of giving their minds and senses to each other and the world.**

These *attentionauts* and *attentionistas* draw on the wisdom of diverse traditions. Across uncharted terrain, emerging practices of joint attention illuminate new horizons of shared political power. Not only power, but beauty, and grace, too.

This is our movement: the free movement of attention in its fullness, freely shared. We call that transformative goodness *ATTENSITY.* Join us in this heightened and heightening glory—or let us join you!

To bring those possibilities to *bloom,* we heed the countless Attention Activists who are already out there, devising new (and revising old) ways of giving their minds and senses to each other and the world.

Attention Activism
takes many forms.
Are you already
an Attention Activist?
Could you become one?
We offer a roll call of
those who embody the
promise of a genuine
movement of
attentional liberation.

T he great promise of Attention Activism is in the astounding fact that its future leaders are out there, *all around us, ALREADY!* Across the country and around the world, people are giving their attention (to each other, to themselves, to their surroundings) in ways that resist commodification and create new possibilities for the sharing of time and space—for the sharing, in sum, of life itself.

Some of these people already identify as Attention Activists. Most, though, don't—or not yet, anyway. But this latter group— folks going about their business, quietly cultivating attention in joyful and life-affirming ways—are walking testament to the fact that, to take down the frackers, *we already have everything that we need.* At the critical moment, our most immediate and useful tool is always the one closest at hand. For this reason, it is essential to take some time to recognize the ways that every one of us is already pushing back against the financialization of our humanity, and, in doing so, to band together with all the others who, in a million different ways, are up to the same thing.

We've already said that Attention Activism is a collective effort. And the prelude to any collective effort is a *roll call.* Who's on our team? How are their skills? What can we do to sync up what we've got?

There are SO MANY! Once one begins to look around, the Attention Activists are *everywhere.* See the little ones dragging their instrument cases to school or weekend practice? They are all on board, and their parents, too! We'll put them together with the casual-but-committed athletes who show up for their games, everyone who is a regular at Tae Kwon Do, however maladroit. Throw in the kids from Chess Club! Plus the HOME COOKS! And maybe the crabbers, bird-watchers, and anyone who has taken a pottery class. These are the **AMATEURS**—a name that contains the Latin word for "love." When we do what we do not because we are the best, but because good things are good to do, and the goodness lies in doing them, and learning how to do them a little better ... *THAT* is attention made active across time. It is a powerful antidote to the quick twitch of the frackers. And these activities make shared communities of IRL endeavor.

But Attention Activists come in all stripes. Take Josh, of Boise, Idaho. Josh is a welder and has been since he discovered the craft in a high school shop class. He recounts with a smile his first time at the bench: the explosion of light, the flying sparks, the heat, and the sputtering hiss of the torch was an "assault on the senses." Josh had always been into tools and handiwork. His happiest childhood memories include playing Bob the Builder with his grandfather, a farmer who, in Josh's fond recollection, could conjure just about anything with a trip to the lumber shop, a patch of dirt, and his two hands. To build on such an elemental level—directly to fuse two pieces of metal—struck him as a near miracle.

Every Attention Activist has a PRACTICE. This is the mode of attention that brings us most intimately into connection with ourselves and the world. For Josh, that magic moment happens when he flips his hood, leans over his workbench, lights his TIG torch, and with his left hand feeds a slender filament of alloy into the hissing heat. Slow, steady circles, he will advise. You've got to get a feel for the metal. Move too quickly and the weld won't take. Too slowly, and the molten puddle will bubble and bunch. But, in his telling, the torch and the metal do all the work.

With the hood down, he is effectively blind—that is, until the torch is firing. Then his mind narrows to the light of the flame (it's like getting sucked into the bright vortex) and all noise fades. What was previously a concrete, goal-oriented task becomes something far stranger—a kind of dance with the luminous puddle of molten metal at the convergence of his mind and senses.

Josh's absorption by the "puddle" is a testament to the unique forms of attention that create our physical environment. And he's not alone. There are countless folks like him who are responsible for the skilled craft and labor that build our bridges, maintain our towns, and keep life-giving supplies moving back and forth across the country. Like...Mike, an arborist on the Missouri public works team who is responsible for the upkeep of over a thousand streetside trees across his county. Anyone with the privilege of riding shotgun as Mike cruises through town will learn to see public space in a new way: not in terms of schools and restaurants, or on-ramps and roundabouts, but in leaf-rot and deadwood, in the tell-tale "scales" on a red maple's bark and in a tilting bole. Or how about Engineer DeLay, out in Ames, Iowa, an OG Friend of Attention whose work running the railroads takes him across the vast expanses of the country's interior? As he puts it, "In my trade, all devices are off." Why? Because he has *29 million pounds of grain* behind him, stretching back across "14,800 feet of freight train." That's nearly *three miles* of locomotive! And it is all the rolling tail to that cone of light piercing the darkness—the engine, and Mr. DeLay, with his eyes on the rails, all night.

Engineer DeLay and folks like him are keepers of an attention that is durational, hypnotic, and, also, perfectly alert. Does *anybody* watch *anything* for as many consecutive hours as a locomotive engineer watches those tracks? Or as truckers watch the road? All the while, operating several tons of high-velocity heavy machinery?

Josh and Mike and Mr. DeLay are what we might call **CRAFTERS** and **OPERATORS**. Their highly skilled forms of durational attention are responsible for the infrastructure that makes society tick. As Attention Activists, they stand for the absolute necessity of a free mind to wield the tools, tend the grounds, and build the railings. They remind us that our world—the one where we live, the one we can touch—is actually *made of attention* just as much as it is made of metal and concrete.

Moving from the wide Midwestern plains to the warehouse parties of Detroit and New York City, we find Attention Activist BachTroy, a Brooklyn-based producer and DJ whose Afrobeat performances beckon ever backward to electronic music's African American roots in his Motor City hometown. BachTroy (whose Attention Activism led him to become a facilitator at the School of Radical Attention) sees the dance floor as the ultimate attentional gathering place. Music has a way of tuning minds to the same frequency, as he will often point out, and research has shown that rhythmic dancing can bring the legion heartbeats of a crowded room into perfect synchrony. Who knows? A drumbeat may well be the oldest attentional technology on the planet.

In an unexpected way, BachTroy's dim, thumping, crowded, synchronized Attention Activism is a lot like the work of Laura, a writer in Portland, Oregon. In 2011, Laura started a public art project to serve the city's unhoused population. It would be a *library,* she imagined, but mobile, nimble, on wheels. It would serve precisely those folks who were most likely to feel unwelcome in the city's public library system. It would function as a kind of mobile "attention sanctuary" protected from the fracking of digital spaces on the one hand and the steady drain of poverty and precarious housing on the other. So Laura tricked out a front-loading Haley Tricycle with a wooden box that opened from the top and from a sliding drawer. She filled the library with her favorite titles, painted STREET BOOKS on the side in elegant *Metropolitain* lettering, and set out pedaling. The people she found sleeping in tents by the Willamette River or sitting on the curb at the highway's edge were eager to read, but more than anything they were eager to talk. So she would stop, hand out library cards, and chat awhile. Soon these people became her friends. Nearly fifteen years later, Laura's still at it, although this time with a fleet of bikes and a team of Street Books volunteers, many of whom were among her first unhoused patrons.

On the surface, BachTroy and Laura provide music and books, respectively. One loud, the other quiet. One kinetic, the other silent. But what they both *really* provide is SHARED EXPERIENCE. Collective attention requires other people as much as it requires a shared object of attention, and these Attention Activists serve up both in equal measure. They are joined in these ranks by imams, neighbors-who-throw-dinner-parties, community organizers, tailgaters, pastors, walking tour guides, and hosts of all stripes. Think of them as the **GATHERERS**: people who create life-giving ways of getting together and sharing attention. They are the people who make us Attention Activists an "us" in the first place.

Or consider Nick, a programmer and coding instructor living in Denver. Nick's path to Attention Activism began when he signed up for a web development program right at the beginning of the 2000s' coding boom. He was there at the start of it all: Facebook, YouTube, and Twitter were hiring, and anyone with decent programming chops was virtually guaranteed a place in an industry that aspired, without irony, to "infinite growth." Nick bounced around, designing apps like Angry Birds and games like *Mass Effect 3* and building software for pharmaceutical companies. He was at the center of the tech world—or, as he began to see it, in the belly of the monster.

In 2016, Nick started teaching design and development at a popular programming boot camp in Denver. It was then, when he was tasked with teaching the next generation of developers the inside techniques of the industry, that he began to question the principles creating an increasingly toxic web environment. He saw the ways tech depleted him and his wife in their personal lives, and the ways it encouraged resentment and vindictiveness in society as a whole. It was hijacking politics. Plus, Nick had two daughters. It didn't even bear thinking about what effect human fracking would have on their well-being.

So he asked himself: *How can I DECONSTRUCT this technology that I myself helped build?* The answer came to him in the very situation that produced the question: He would teach his young programming students how to hack the deceptive designs and quick-shill principles that had helped build the modern internet. He would combine his coding sessions with critical histories of the internet's privatization and the implications of an advertising business model. He would teach his students to recognize dark-pattern tricks—design techniques that manipulate users' behavior. He hopes the next generation of programmers can right the wrongs that his generation brought about. And not just that: Nick began to imagine again all the good things these systems could actually achieve. What if, he asks, these powerful tools were geared toward connection and well-being, rather than advertising and corporate profit? What if the internet could be its own *attention sanctuary*?

We can think of Nick and his ilk as the **COUNTER-CODERS** and **MONKEYWRENCHERS** in our vast coalition: the skilled programmers working to reclaim our tools from the interests of the frackers; the folks who know how to tweak and turn and game and invert the attention economy's basic weapons. Their work reminds us that the only worthwhile technologies are the ones that create the conditions for human flourishing.

Roll call! All present: **AMATEURS, CRAFTERS, OPERA-**

TORS, GATHERERS, COUNTER-CODERS, MONKEY-WRENCHERS . . . But who else is out there? Probably we should count the **GAMERS** and **PLAYERS,** too—those who rally with their crew for poker or golf or axe throwing, or who just stay at home to go head-to-head with their spouse on the weekend crossword. Foosball, anyone? And we won't omit those who build worlds together in *Minecraft* or train up as Dungeon Masters. Are these activities frackable? They certainly are! But what matters is what can be done with what we have, and like the clownfish that have learned to live inside the tendrils of the anemones that would eat them, the Attention Activists who learn to thrive inside the poison arms of the beast are a lesson for us all!

Let's change scenes. What about Vivian, who is a lead operating room nurse in the Mount Sinai hospital system in midtown Manhattan? Her job is to keep her eyes on her patients, in a way that is fundamentally different from the surgeons (or, for that matter, the anesthesiologists). The latter are looking at a set of readouts, or are hyper-focused on a small incision. But as another hour-long drill unfolds around the semiconscious body on the table, she is the one who is responsible for the *person* who is working to stay calm across this uncanny and frightening situation. She checks in. She adjusts a bolster. She speaks reassuringly—and, all the while, she is also keeping track of what her colleagues need; she is attuned to the room. Hers is the live attention of actual care, the readiness of those who work to minimize suffering and predict the needs of the flesh (and the spirit). Anyone who has ever received this kind of gift knows just how precious such attention can be—particularly when we are vulnerable. And when are we not vulnerable?

There is a beautiful Greek word we will roll out for these attentional *caregivers*—the therapists, midwives, pastors, the death doulas, the attendants, and all of those PARENTS out there. Everyone who *lays on hands*. Everyone who brings the powerful medicine of actually *listening*. They are the **EPIMELETES,** from ἐπιμέλεια (*epimeleia*), meaning "care, concern, solicitous and nurturing and healing attention."

Then there are those who cultivate special forms of attention to our shared mother tongues—to language itself. Not just the poets and storytellers and translators and stand-up comics (though they all count!). What about the well-scrubbed young'uns of the *National Bible Bee,* a Christian competition where the pre-teen competitors memorize nearly six hundred Bible verses and recite them in front of a panel of judges? Or likewise for the youth participants in the annual USA Quran Competition (Houston, Louisville, Chicago, Maryland)? It's hard to imagine anybody *on Earth* who is giving as much sustained attention to texts (ancient texts, no less!) as these youthful orators. Add to their company the scattered enthusiasts of Emily Dickinson and those backseat prodigies boasting an encyclopedic recall of Lil Wayne's labyrinthine and tongue-tripping rap canon, and you have a network of Attention Activists with a precious and endangered command of the power of spoken language: the Orthodox cantors, the slam poets, the rappers, those who steep themselves in ancient texts and become living mouthpieces for rich traditions. Call them our **BARDS** and **RECITERS.**

And what about the faithful few who *stand for hours in the bone-deep predawn cold along the high-altitude route of the Leadville Trail 100 ultra-marathon*? We aren't talking about the runners, now (stick them in "Amateurs," above). The world of ultra-running (where competitors pound out the mileage equivalent of back-to-back-to-back marathons) is populated with FANS like this, who line the 100 miles of alpine track to cheer on the increasingly haggard runners filing past. There are no medals for these stalwarts who pass out orange slices and energy goo; their only comforts are thermoses of hot tea and sleeping bags. And yet they stand there all night, a shivering testament to the ways that attention can make otherwise excruciating feats into something triumphant, even fun.

These cheering onlookers have fellows far and wide: in the stands at Little League games, on skating rink bleachers, in elementary school auditoriums, and clustered around stages in leaky bar basements from Cleveland to San Francisco to Tampa. What connects these people is the uplifting power of their joint attention. They are there to watch and listen and cheer for somebody they love. Anyone who has ever risen to greatness atop a mound or a stage has done so because of the generous and exuberant attention of people like these. What do we call them? Easy! We call those people **SPECTATORS**—and beyond them (in the Reddit threads and fan-fic subcultures) we see the wider worlds of the **FAN-DOMS**. These are the well-staffed armies of those whose love and care manifest as time and presence and *entanglement*—with their people, their games, their stars!

And then, of course, there are countless kinds of attention to which we have, as yet, given no name. What about the guy who stands in the parking lot outside baseball stadiums hoping against hope that a home-run ball will clear the upper rows and fall into his outstretched mitt? What kind of attention is that? And what can *he* teach us? We don't know—not yet, at least. But we're willing to bet that, sooner or later, we may learn the lessons of such an improbable pastime.

Or perhaps the answer is right there before us: the value of the curbside catcher *is in the curbside catching itself*! Which is to say that, for this guy, the experience of giving his attention for hours with little promise of reward is justification enough. This may appear perfectly obtuse to many, but it contains, in fact, a deep wisdom: that there is a kind of value that cannot be assessed by translation into something else. Attention creates this kind of value in the lives of many, many people; it is a source of fulfillment that needs no justification. Goods such as this are rare indeed. Their preciousness is in the way that they puzzle us, and please us all the while. In this spirit, let's make space for those whose attentional commitments confuse and provoke us: Let us call them our **PUZZLERS**.

Roll call: ongoing! For our ranks are growing every day. Not only with "converts" to the cause, but also with good people everywhere who come to realize that they are *already* Attention Activists *par excellence*—that they have been fighting the good fight all along! They have the tools, and come to heft them, to share them, to feel them doing the powerful work those tools can do. Maybe you identify with one, or several, of these groups. Or better yet, perhaps you can think of a posse that we have overlooked. In any case, this rough-and-ready roll call is hardly exhaustive; it's more *invitational.*

Where in your life do you feel in control of the movement of your attention? What leaves you feeling better, not worse, after you give your mind, time, and senses to it? When do you notice an experience of your attention leaving you feeling curious, energized, *replenished*? When do you feel closer to the reality of the world and other people? And where does your attention move in ways that can't be monetized by the human frackers?

Can you build on it? Share it with others? (And elude, sabotage, or double-cross the frackers, if they try to worm their way in?)

If you are out there (and we know you are), then join us! Teach *us* to attend in the ways that feed your soul and fill your days with meaning. There is so much to learn, and the best way to learn is to do it TOGETHER.

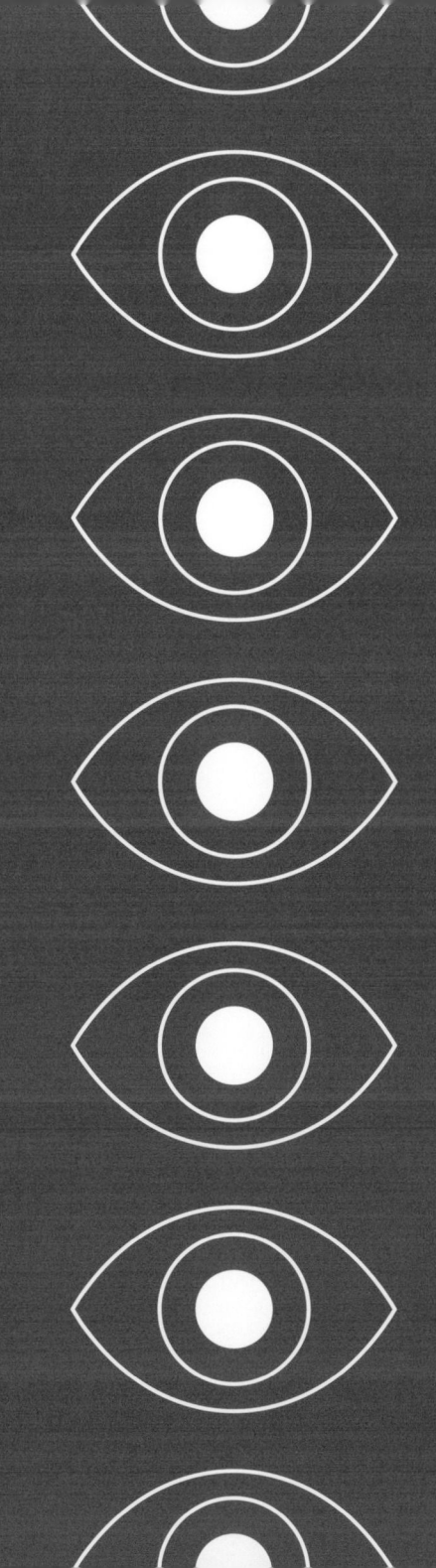

ATTENSITY!

A Manifesto of the Attention Liberation Movement

You are correct: Something is seriously wrong. It has to do with our ATTENTION, our essential ability to give our minds and senses to the world. This precious capacity has been channeled, captured, and commodified by an industry of immense technological and financial power. How? Call it "human fracking."

Human fracking is bad for people, and for politics. It reduces our very beings (and our relationships) to that which can be quantified, bought, and sold. All this is the triumph of a catastrophic *lie* about what it means to be human. But deceit and exploitation are never inevitable. To push back, we need more than isolated, individual efforts; what we need is a movement of collective resistance.

This movement of attentional liberation exists and has a name: ATTENTION ACTIVISM.

Attention Activism is a fight for justice. This emancipatory uprising takes our apocalyptic present, turns it on its

head, and creates, from the chaos and confusion, new vistas of human flourishing.

Attention Activism is rooted in STUDY—a commitment to diverse forms of teaching and learning centered on attention (what it is, what it can be, what it can do). Attention Activism also requires COALITION-BUILDING—collaboration and solidarity across a range of communities who see attention's essential role in human flourishing. Finally, Attention Activism means the formation of SANCTUARIES—spaces where people can gather, care for each other, experiment with different kinds of attention, and conceive brighter futures.

To discern the revolutionary possibilities of the present, we look to artists, thinkers, and dreamers. To bring those possibilities to bloom, we heed the countless Attention Activists who are already out there, devising new (and revising old) ways of giving their minds and senses to each other and the world.

These *attentionauts* and *attentionistas* draw on the wisdom of diverse traditions. Across uncharted terrain, emerging practices of joint attention illuminate new horizons of shared political power. Not only power, but beauty, and grace, too.

This is our movement: the free movement of attention in its fullness, freely shared. We call that transformative goodness *ATTENSITY*. Join us in this heightened and heightening glory—or let us join you!

These *attentionauts* and *attentionistas* draw on the wisdom of diverse traditions.

Viewed correctly, sources of resistance to the human frackers can be found everywhere. Figure out what you've got. Deepen it. Then share it.

Situations change everything. This is one of the basic truths. And it is, in its essence, a truth about history. Things mean different things in different moments. This is fundamental. A steel-nibbed fountain pen was the *coolest new gadget in the universe* in 1820. In 1920, it was just an ordinary doodad. By 2020, it indicated that you were some kind of weirdo. Time matters. Looked at one way, that is kind of an obvious thing to say. Looked at another way, it's one of the most profound insights available to humans.

Changing times change the meanings and uses of everything. Let's take an example. Pitchforks are useful for pitching hay. Shovels are good for digging holes. An axe is ideally suited to cutting down trees. And a large iron pot with a wooden spoon would normally be used for making soup—soup that one might wish to eat with others while sitting in chairs around a table. HOWEVER, in April of 1834, angry citizens of the French city of Lyon carried their tables, chairs, pitchforks, shovels, axes, soup pots, and assorted other hardware and furniture down into the streets to erect barricades, and arm themselves for resistance against agents of the crown whose exploitative practices had finally driven people to defend themselves.

It's a dramatic image, but not an absurd one to invoke in our own moment. For the Attention Activist it is a reminder that our *inherited tools*—what we have to hand, what we can take up and use—are our arsenal as we get serious about defending ourselves against the frackers. What can we grab, and take into this nonviolent fight?

That isn't a question we can answer for you. It is a question you need to answer for yourself—and for us! We need a new kind of inventory, as we go over what we know, what we have stashed away, what *someone we know might know*—and we go get it, and BRING IT OUT.

That soup pot reminds us of one of our favorite old stories of catalytic community change: the tale of the trickster-feast. Sometimes remembered as "Stone Soup" (or "Axe Soup," or "Nail Soup," or "The Legend of the Kamper Onion," etc., etc.). You probably know the way it goes—at least in one of its hundreds of versions, which have circulated as folk tales across Eurasia for as long as anyone can remember. One or more hungry characters show up in a village that is absolutely unwilling to give a stranger a meal—food is tight, everyone is looking out for themselves, the doors are bolted, and nobody wants to come outside. "Okay," announces the stranger, nice and loud, "I guess I'll have to make STONE SOUP!"

And the locals are intrigued. Stone soup? Is that a thing?

"Yeah, yeah, it's totally a thing! All I need is a pot, and we're off to the races! I've got a few stones right here . . ."

And so begins a tale of gentle and philanthropic *beguiling*. Since the boiling of a few rocks won't itself make much of a soup, it is true. But the *idea* of Stone Soup is so compelling that it gradually creates the conditions for collective collaboration and mutual aid. A few noses poke out of their doors, as the stranger stirs the pot. "Yep, it's gonna be a fine stew, since these are pretty good rocks you have around here; though, of course, it would be a *wee bit tastier* if we had an ONION . . ."

And sure enough, a few onions appear, as a contribution to the crazy project. And someone else brings a bunch of carrots up out of the cellar. Before long, everyone wants to put something into the pot, and a giant cauldron of *actual soup* is indeed eventually made . . . out of the curiosity and emerging goodwill of the community. By evening, the village is out in force for a magical banquet, feasting on the miracle of their own willingness to *bring what they had*. And share.

Are you worried about that initial "fib"? The thing about *making stone soup*?

That's legit. In 1971, Hannah Arendt published an essay titled "Lying in Politics," which has been much anthologized since. The essay is about a lot of things, but at its heart is an observation about the complex relationship between visionary politics and tell-it-like-it-is truth. After all, among the things we need to live well is a sense of something *better,* something *out ahead*—toward which we can move, together, in hope and aspiration. Such visions (of promise and possibility) aren't really issues of truth or falsehood, exactly. They are, rather, the work of imagination. They are *dreams*. By these lights, "stone soup" isn't really a "lie"—it's more like one of those very special stories that the literary theorists call a "hyperstition": a tale that has the power to make itself true. Some of these are bad, and some of these are good. But one doesn't evaluate them on the basis of the conformity with the world-as-it-is. One evaluates them on the basis of the worlds they *bring into being.*

There is no culture without such promises. But in the attention economy, we have ceded the writing of these scripts to those who would sell us to ourselves, for the price of our souls.

It isn't a good deal, and as Attention Activists we decline to take it. Instead, we will go find other stories, and bring up and share other ways of dreaming. What we have, what we need, are all the very different traditions and practices of human attention: techniques of meditation and prayer, modes of athletic concentration and focus, forms of spiritual leisure and play, the solicitous eye of mothers and grandmothers, the street smarts of the bona fide hustler, the essential absorption of a child in a sandbox. These enterprises are as different as pitchforks and soup ladles! What does Vipassana meditation (durational attention to the breath passing over one's upper lip) have to do with that youngster in a *batting cage,* trying to learn how to hit a sinker? And what do *either* of these have to do with my neighbor, blithely walking her Yorkshire terrier?

Under ordinary circumstances, this seems like a perfectly random assortment of goings-on. But under the perfectly *non-ordinary* conditions of our collective confrontation with human fracking, each of these activities becomes a valuable part of what we all need to bring to the barricades, what we need to bring to the banquet we'll share once we've walled out the dark artists of spirit-commodification!

These new conditions reactivate our past and remake our inheritance. We see old things anew. Are you a little embarrassed about your evangelical cousin? The one whose kid is being forced to learn the Bible by heart? Think again! There is surely something very special in that form of attention-formation. Anxious about the lingering gender stigma of domestic labor? And the complex shame that too often comes with both doing and *not doing* childcare? And what about that "Great Books" curriculum at the college on the hill? Again, there is a world of rethinking to be done, as those complicated legacies are reconceived in attentional terms. How can those traditions be *renewed,* and serve *all of us* differently as we fight back against the frackers? Everything we have, re-seen through the lens of attention, becomes a way forward: music, sports, cooking, Talmud, feeding the baby, sitting with those who are soon to die.

Grab up the attentional practices you've got! Have your neighbor teach you what she knows! And share what you're learning! Because this is literally *all we have.*

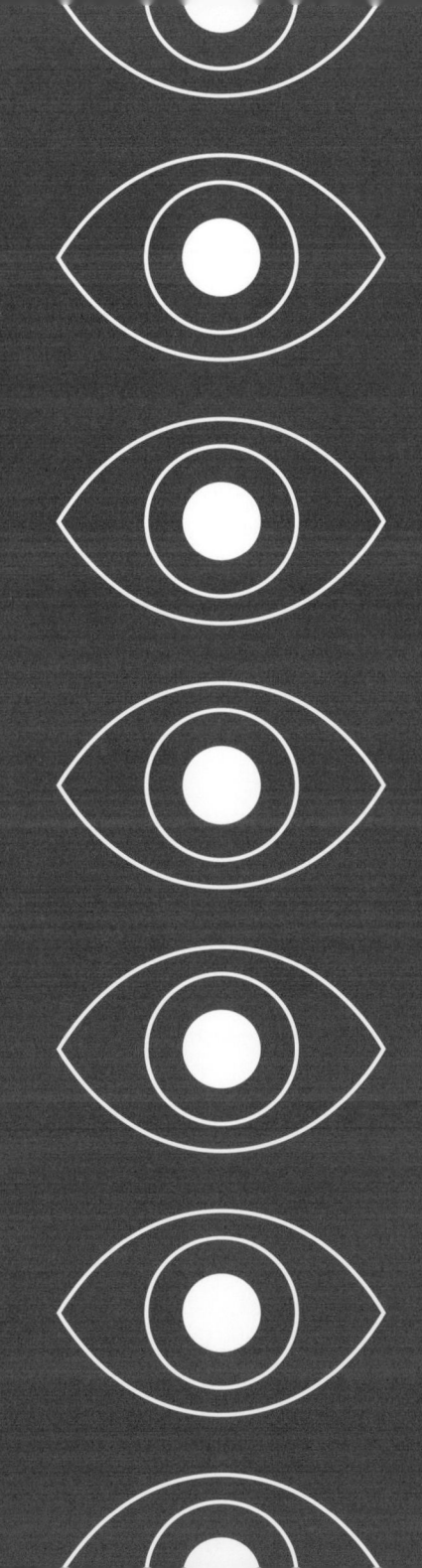

ATTENSITY!

A Manifesto of the Attention Liberation Movement

You are correct: Something is seriously wrong. It has to do with our ATTENTION, our essential ability to give our minds and senses to the world. This precious capacity has been channeled, captured, and commodified by an industry of immense technological and financial power. How? Call it "human fracking."

Human fracking is bad for people, and for politics. It reduces our very beings (and our relationships) to that which can be quantified, bought, and sold. All this is the triumph of a catastrophic *lie* about what it means to be human. But deceit and exploitation are never inevitable. To push back, we need more than isolated, individual efforts; what we need is a movement of collective resistance.

This movement of attentional liberation exists and has a name: ATTENTION ACTIVISM.

Attention Activism is a fight for justice. This emancipatory uprising takes our apocalyptic present, turns it on its

head, and creates, from the chaos and confusion, new vistas of human flourishing.

Attention Activism is rooted in STUDY—a commitment to diverse forms of teaching and learning centered on attention (what it is, what it can be, what it can do). Attention Activism also requires COALITION-BUILDING—collaboration and solidarity across a range of communities who see attention's essential role in human flourishing. Finally, Attention Activism means the formation of SANCTUARIES—spaces where people can gather, care for each other, experiment with different kinds of attention, and conceive brighter futures.

To discern the revolutionary possibilities of the present, we look to artists, thinkers, and dreamers. To bring those possibilities to bloom, we heed the countless Attention Activists who are already out there, devising new (and revising old) ways of giving their minds and senses to each other and the world.

These *attentionauts* and *attentionistas* draw on the wisdom of diverse traditions. **Across uncharted terrain, emerging practices of joint attention illuminate new horizons of shared political power.** Not only power, but beauty, and grace, too.

This is our movement: the free movement of attention in its fullness, freely shared. We call that transformative goodness *ATTENSITY*. Join us in this heightened and heightening glory—or let us join you!

Across uncharted terrain, emerging practices of joint attention illuminate new horizons of shared political power.

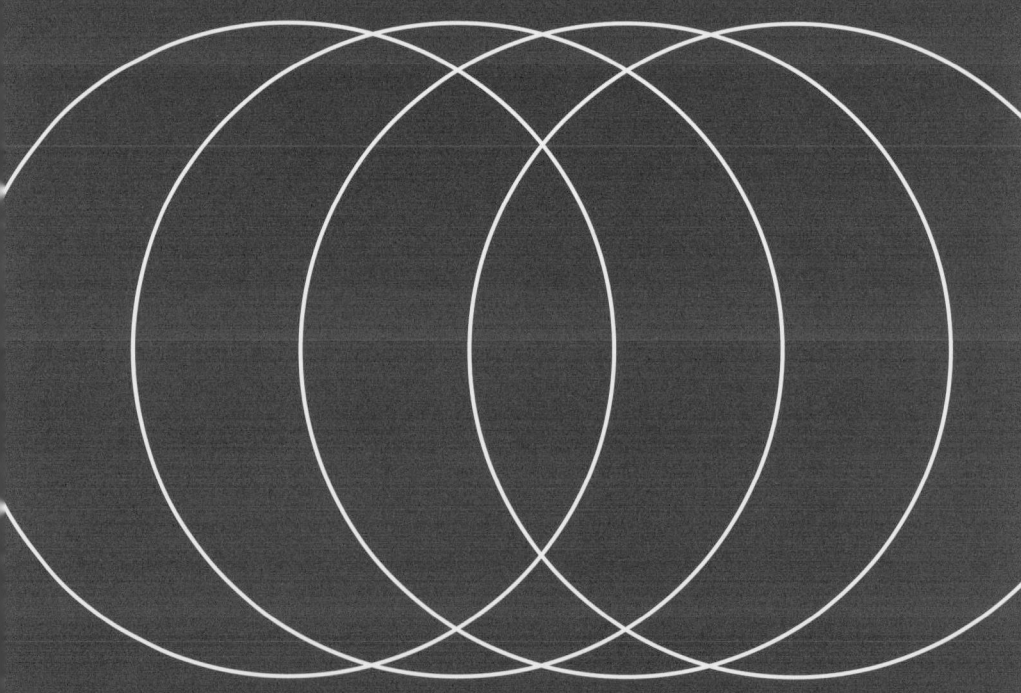

New forms of exploitation invite new forms of political action, and create new political situations. With imagination and will, we can turn our shared attentional condition into a new kind of collective identity—and power.

Revolutionary change is driven by political power on every scale, from grassroots, park-bench advocacy to the echoey antechambers of legislative reform. The Friends of Attention tend to the more intimate side of this work. At the School of Radical Attention we have built out a curriculum of courses, free workshops, and meetup street-pedagogy that centers on attention—and aims at cultivating actual community. We support teachers who want to pilot attention-centered activities in their classrooms, and we have begun extending a program of fellows and allies and "study groups" across the country. We write, we rabble-rouse, we *DO STUFF.* We explore attention together, and we talk about that work to anyone who will listen. Any meaningful transformation of our society, we believe, requires broad networks of care and solidarity, as well as so-called mobilization structures to channel the energy of the populace toward meaningful political action. It's our business to build these coalitions, and, in the words of adrienne maree brown, "to move at the speed of trust." To that end, we are always feeling around at the edges of what is recognizably "political"—for we believe that politics, like everything else, changes with time, and that the present moment brings changes that we are only beginning to understand.

That said, we are political realists, and we cover all our bases! Our eyes are ever on the hallowed halls of the highest worldly powers, and we follow changes in that realm with intense interest. And there, too, we find bad news and good news. The bad news is that the frackers have achieved a scope and scale of power that is nearly unprecedented in world history. Their chokehold extends from congresspeople to eleven-year-old children. The *good* news, though, is that the past few years have seen promising signs of legislative and policy initiatives that fight back against the pump-and-suck impunity of attentional exploitation.

These are welcome advances for Attention Activism. No movement can change the world without turning the great machinery of political institutions. But it is *also* a defining spirit of our work that we choose not to take any load-bearing words or ideas for granted. This applies to the category of "politics." As it turns out, there's more to politics than what happens in statehouses! Part of our work is to tease out the ways that *politics itself* is being reshaped and redefined by the human frackers. Indeed, at the very HEART of Attention Activism is the idea that attention has become "political" in historically novel ways. That word—*political*—is a popular one that serves lots and lots of different purposes. It can also be stretched so thin as to become transparent and mean nothing at all. So let us be concrete about what, exactly, we mean by the "politics of attention."

There are a few levels to our claim that *attention is political*. The **FIRST** is that "politics"—what your aunt or uncle probably means when they use the word—has always been about attention! Name the first hot-button topic that comes to mind: immigration, welfare, reproductive rights, voting rights, what have you. Each of those topics became a political "issue" because enough of the American public decided it was worthy of their attention, at which point people began to disagree over what, exactly, to do about it. This disagreement—channeled, mediated, and (in the best cases) resolved through a set of deliberative processes—is what we call "politics." It is what you find when you tune into C-SPAN.

In other words, attention is what you might call a *precondition of politics*. We need a certain degree of collective attention (attention to the same things, and attention to each other) for the creaking gears of state to start turning.

But collective attention does not just happen; it is constructed. This requires work—the reality is that most problems haven't even made their way to the ballot, or to the capitol halls. In these cases, it is the work of advocates and activists to bring their issue to the public attention, and to insist upon it until enough people care to start changing things.

This is what we see when we look backward at the long history of movement politics. Again and again, the first step of any social or political transformation has been a group of people insisting that the thing they care about *demands* collective attention. Abolition! Suffrage! The environment! Nuclear disarmament! The harms of Big Tobacco! This is what Attention Activists across the country and the world are doing *right now*. They are saying, as we have said in these pages: *Hey! Notice that thing you call your attention? Something's not quite right, eh? Well, you're not alone—just about EV-ERYONE feels this way. And it's no mystery why that's the case: because the same system that is fracking your attention is fracking mine, too. We're both affected by this problem. Which means that we can work together to solve it.* This is the kind of work that is pushing for pioneering legislation (like the Stop Addictive Feeds Exploitation Act, or SAFE, in New York State, for instance; or the ongoing fight in Congress for some version of the Kids Online Safety Act, which passed in the Senate in 2024).

What these activists and advocates are doing, in other words, is calling attention to attention itself as a matter of urgent collective concern—and that's the **SECOND** way in which, right now particularly, *attention IS political.*

And this is critical. Only when people from all walks of life have come to recognize, counterintuitively, that the condition of their innermost selves is a matter of public importance—only then can attention enter the sphere of what is traditionally meant by the idea of *politics.*

But these first two ways that "attention is political" still both use the *same* basic meaning of the word "politics." The **THIRD** way that attention is political relies upon a concept of *POLITICS* that is less familiar, and deeper. It requires a higher dose of imagination, and a bit of historical thinking.

Recall our earlier account of the Industrial Revolution: how, in the nineteenth century, new technologies produced the ability to concentrate, optimize, and extract physical labor at a scale never before seen. The awesome powers of this novel system were matched by the brutal conditions of exploitation that it imposed upon a new class of workers, who were defined, suddenly, by the very thing that nascent industrial capitalism sought to take from them: their labor. Gone was the deep web of customs and community, of relation to land and tradecraft. Appearing in its place was the increasingly pure logic of a *market economy,* where every person appeared, first and foremost, as an economic actor whose every minute of physical exertion could be valuated and sold.

The Industrial Revolution refers to a transformation in what historians call "modes of production"—the way that things (and values) are *made* by a society. We see, in retrospect, how this new mode of production, built in turn upon new zones of expropriation (the exertions of the body, systematized and harnessed by machines), produced a new kind of being. The levers and flywheels of the great industrial mills didn't simply "transform" the cotton bales that emerged from the factories as bolts of cloth—they also *refigured* the social formations of the people who worked within the factories. A worker, looking to her left, would see another WORKER. What people had in common was *the very thing that was being taken from them.*

One result of this staggering change was the formation of a new kind of power: the power of workers *collectively to WITH-HOLD their labor.* This, in turn, created a new kind of politics, and new kinds of communities, known as unions, that enacted those politics.

Think, for a moment, of the feudal system from which industrial capitalism emerged: Back then, there was no such thing as a peasant *strike*. Feudal peasants could not "paralyze the machine" of feudal power because feudal power was *not mediated by machines.* Sure, peasants could *revolt.* But the food they refused to cultivate was the same food they then didn't get to *eat.* This limited their leverage. But it was only with industrialization that the refusal to work, exercised independently of violence (torches, pitchforks, slave revolts), became a genuinely powerful way to push back against oppression. When the factories stopped churning, it placed the owners in a vise.

Things often look obvious in retrospect. But we must remember that, at the beginning of this historical shift in the mid-eighteenth century, the people of Manchester, Leeds, and Liverpool could not have imagined the political future that awaited them. Previously unknown kinds of violence lay in store—but so, too, did previously impossible points of leverage, and new kinds of collective power.

We—the Friends of Attention, and many of our accomplice networks, too; all those committed to the *Attention Liberation Movement*—believe that the present moment constitutes a similar watershed. Whereas the industrial economy enabled a new and wholesale extraction of our physical capacities (the machinic exploitation of the movements of the body), the emergence of human fracking has enabled a new and wholesale extraction of our *attentional* capacities (the exploitation of the movements of the mind and senses). And like the workers at the dawn of that new epoch, we have reason to believe that there are unknown forms of power before us, and unknown kinds of politics.

So . . . what is it we see? We are at the threshold of tremendous changes. Predictions are not easy. But it is clear enough that a historical shift of this magnitude will leave nothing untouched—including our very sense of ourselves, and the kinds of political AGENCY actually available to us.

Even this has happened before. Early industrial capitalism in England dissolved ancient ties of reciprocity and relation, stripping away complex social worlds and rendering the English people as new subjects defined by their participation in a market economy: *Homo economicus.* Might our present moment of dissolution leave us freshly subjectivized? Do we stand before each other as a new kind of political being—as ***Homo attentus***?

We suspect so. And *Homo attentus* will be a new kind of creature—capable of giving shape to new worlds through the collective enactment of new kinds of attentional freedom. The very *nature* of politics will be transformed.

Indeed, we believe that, in the face of an increasingly squalid and brutal traducing of democratic ideals, this new THIRD kind of attentional politics—*committed to the authentic FREEDOM of attention and what it can do*—prefigures the overthrow of what has rightly been decried as the "society of the spectacle." That monstrous system of media, money, and falsehood that increasingly keeps us enthralled to a cacophonous circus of illusions.

We call this *newly political* attention . . . ATTENSITY!

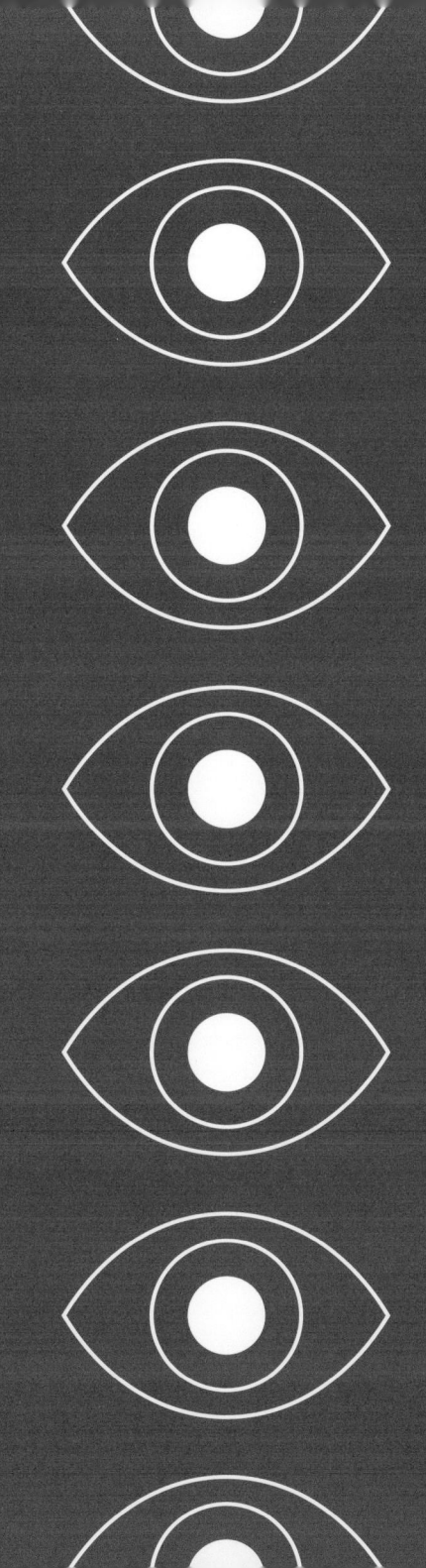

ATTENSITY!

A Manifesto of the Attention Liberation Movement

You are correct: Something is seriously wrong. It has to do with our ATTENTION, our essential ability to give our minds and senses to the world. This precious capacity has been channeled, captured, and commodified by an industry of immense technological and financial power. How? Call it "human fracking."

Human fracking is bad for people, and for politics. It reduces our very beings (and our relationships) to that which can be quantified, bought, and sold. All this is the triumph of a catastrophic *lie* about what it means to be human. But deceit and exploitation are never inevitable. To push back, we need more than isolated, individual efforts; what we need is a movement of collective resistance.

This movement of attentional liberation exists and has a name: ATTENTION ACTIVISM.

Attention Activism is a fight for justice. This emancipatory uprising takes our apocalyptic present, turns it on its

head, and creates, from the chaos and confusion, new vistas of human flourishing.

Attention Activism is rooted in STUDY—a commitment to diverse forms of teaching and learning centered on attention (what it is, what it can be, what it can do). Attention Activism also requires COALITION-BUILDING— collaboration and solidarity across a range of communities who see attention's essential role in human flourishing. Finally, Attention Activism means the formation of SANCTUARIES—spaces where people can gather, care for each other, experiment with different kinds of attention, and conceive brighter futures.

To discern the revolutionary possibilities of the present, we look to artists, thinkers, and dreamers. To bring those possibilities to bloom, we heed the countless Attention Activists who are already out there, devising new (and revising old) ways of giving their minds and senses to each other and the world.

These *attentionauts* and *attentionistas* draw on the wisdom of diverse traditions. Across uncharted terrain, emerging practices of joint attention illuminate new horizons of shared political power. **Not only power, but beauty, and grace, too.**

This is our movement: the free movement of attention in its fullness, freely shared. We call that transformative goodness *ATTENSITY*. Join us in this heightened and heightening glory—or let us join you!

Not only power, but beauty, and grace, too.

Everything flowers, from within, of self-blessing. Though sometimes it is necessary to reteach the world its loveliness.

n a beautiful poem entitled "Saint Francis and the Sow," the American poet Galway Kinnell imagines an encounter between the legendary animal-loving saint and a somewhat-the-worse-for-wear fat pig, lying in a stinking sty and nursing a squirm of muddy piglets. There isn't much action in the poem. All that happens is that the saint comes over and puts his hand on the sow's head. That's it, really. Except, somehow, this small gesture is depicted, in the poem, as effecting some sort of nameless current of vitalizing glory that tingles all through the scene—a current that runs through the body of the broken animal, through her brood, and, by extension, through the world itself. The current is described as a "remembering" (of the sow, by the sow) of . . . SOW! "Sowness" is what is recalled, and made *magnificent,* in this act of touch that is also an act of mutual recollection. The sow, and the saint, and we, the readers, become inward, for a moment, with "the long, perfect loveliness of sow," as the final line of the poem puts it.

What has happened?

This poem launches with something like a *theory of everything,* which it states in its opening lines: "The bud / stands for all things." Which is to say, it is the contention of the poem that everything in the world is properly understood as "bud-like"—meaning, everything in the world is a *flower waiting to "happen";* waiting, that is, to open into its florescence. And Kinnell (or is it Saint Francis?) offers a very concrete account of what causes the bud of things to burst into full blossom: Even "those things that don't flower" (in any botanical sense) actually CAN, we are told, *"for everything flowers, from within, of self-blessing."*

If this is the case, though, then perhaps there is nothing to be done? We need only step back, and the beings of the universe (perhaps even the objects, too) will simply go about their mystical business of blooming-to-themselves.

But this is not, in the end, what the poem asserts. While it is true that "everything flowers, from within, of self-blessing," it is ALSO the case that, we are told, "sometimes it is necessary / to reteach a thing its loveliness."

This is the exquisite moment at the heart of this beautiful and important poem: The world flowers, and the world flowers on account of its native graces; and yet, sometimes, it is indeed necessary to "reteach a thing its loveliness." It is this that Saint Francis effects in the poem, as he lays his hand on the big, sad, ill-fated pig, and in that gesture causes the beautiful work of self-blessing to begin to work inside the "long, perfect loveliness of sow." The gesture is the catalyst of an inner unfolding—a kind of strong spring sun that makes the buds uncoil into their full and open offering.

"Saint Francis and the Sow" is a poem about attention. Indeed, it is best read as an allegory of radical human attention, since it is precisely *radical human attention* that is the way we "reteach a thing its loveliness"—and in so doing, cause its bud-nature to flower forth. Is this mysticism? We would say no. A mystification? Not at all. A *mystery*? Sure. If you like. But that doesn't make it wrong, or "fake," or "just talk."

That we can *feel* the space between ourselves and the stars— this, too, is a mystery. But life is made of such things. Truly.

There is a technical term in botany for the process by which a bud comes into flower: "anthesis." *Coming into flower* means something specific to the plant scientists. It means the "maturity" of the organs by which the plant can reproduce. It means that the plant is ready to pollinate and be pollinated, to *seed forth*. Anthesis means, then, the irreversible opening into the phase of dynamic beauty that is also a phase of readiness to come into communion with others—and in so doing, to expand more fully into the world. The Friends of Attention have come to use this lovely word, anthesis, as a kind of peaceful battle cry. It is what happens when attention re-teaches us our loveliness, and in so doing, causes us to come into flower . . . *together.*

ATTENSITY!

A Manifesto of the Attention Liberation Movement

You are correct: Something is seriously wrong. It has to do with our ATTENTION, our essential ability to give our minds and senses to the world. This precious capacity has been channeled, captured, and commodified by an industry of immense technological and financial power. How? Call it "human fracking."

Human fracking is bad for people, and for politics. It reduces our very beings (and our relationships) to that which can be quantified, bought, and sold. All this is the triumph of a catastrophic *lie* about what it means to be human. But deceit and exploitation are never inevitable. To push back, we need more than isolated, individual efforts; what we need is a movement of collective resistance.

This movement of attentional liberation exists and has a name: ATTENTION ACTIVISM.

Attention Activism is a fight for justice. This emancipatory uprising takes our apocalyptic present, turns it on its

head, and creates, from the chaos and confusion, new vistas of human flourishing.

Attention Activism is rooted in STUDY—a commitment to diverse forms of teaching and learning centered on attention (what it is, what it can be, what it can do). Attention Activism also requires COALITION-BUILDING—collaboration and solidarity across a range of communities who see attention's essential role in human flourishing. Finally, Attention Activism means the formation of SANCTUARIES—spaces where people can gather, care for each other, experiment with different kinds of attention, and conceive brighter futures.

To discern the revolutionary possibilities of the present, we look to artists, thinkers, and dreamers. To bring those possibilities to bloom, we heed the countless Attention Activists who are already out there, devising new (and revising old) ways of giving their minds and senses to each other and the world.

These *attentionauts* and *attentionistas* draw on the wisdom of diverse traditions. Across uncharted terrain, emerging practices of joint attention illuminate new horizons of shared political power. Not only power, but beauty, and grace, too.

This is our movement: the free movement of attention in its fullness, freely shared. We call that transformative goodness *ATTENSITY.* Join us in this heightened and heightening glory—or let us join you!

This is our movement: the free movement of attention in its fullness, freely shared. We call that transformative goodness *ATTENSITY.* Join us in this heightened and heightening glory—or let us join you!

Don't worry about your "attention span." Worry about reclaiming the deeper, broader, and more various forms of attention at the center of curiosity, play, love, and freedom. From the shared experiences of these other attentional modes will arise a new and better world.

t took a series of harrowing environmental disasters—and the visionary work of thought leaders like Rachel Carson (author of *Silent Spring*)—to reveal the web of life that makes the environment a recognizably *collective*, and therefore political, good. This happened in the late twentieth century.

Now we are well into the twenty-first century, and a comparable process of industrialized exploitation-plus-destruction is taking place. Only this time, it's not water or air that is being despoiled, but the intangible stuff of the human spirit: our curiosity, our hopes, our relationships, our desires, our aspirations, and our actual *dreams*. These are the movements of the mind that power the mills of the attention economy. And these are the parts of ourselves that we are coming to see not as private pools (for self-gazing), but a river from which all of us draw water. The activating insight of the environmental movement was that all bodies are materially connected, and that, therefore, so is our physical health. The attention movement is teaching us a parallel lesson: that our *attention* is connected. That our reaching and receiving and seeking all move through the same systems. That we are in this world *together*. That what we do with our minds and our time and our senses is what gives us what there actually is—the experience of being here.

All this—the immaterial stuff of our being, on which the world literally depends—needs protecting. It needs a movement akin to the environmentalism of the 1960s and '70s. Because the goodness of consciousness itself is now being remorselessly laid waste by a shocking new industry that recklessly and efficiently and surreptitiously converts human care (and interest and desire) into... MONEY. And this deranged enterprise is harming us all.

Attention Activism says NO! The Attention Liberation Movement *pushes back* against the damage.

But we don't just "resist" the bad stuff. We are after much more than that! We are continuously working, in these pages and in everything we do (at the School of Radical Attention, in our alliances and coalition-building and advocacy and listening), to offer an affirmative vision of something BETTER—*a world remade through attentional emancipation.* Our motto for this bold campaign? "ATTENSITY!"

A few words about the word, and the work we hope it can do. We like to think of it in parallel with the term "ecology." Nowadays, that's a term that invokes a whole *universe* of aspiration. It functions as a kind of shorthand for nothing less than a worldview, one that stresses interspecies connectedness, planetary limits, and environmental sustainability. How did seven letters come to bear such a heavy symbolic load? After all, the word was hardly known (outside an eccentric subset of zoologists) until about 1960! It had weird origins (in nineteenth-century Germany, among some slightly creepy dudes), and mostly dealt with technical ideas about predator-prey relationships. But by 1970, ECOLOGY NOW! was emblazoned on flags and T-shirts and became the battle cry of a youth culture committed to utopian revolution. The term had become a way of saying *STOP DESTROYING THE PLANET!* And it meant *WE WANT SOMETHING BETTER FOR OURSELVES AND OUR CHILDREN!* It had its origins in science, but it broke out as a slogan for participatory, radical *renewal*— across the domains of nature, culture, politics, and individual life.

We call for ATTENSITY in this spirit! Like "ecology," our term hails from the sciences. Back at the start of the twentieth century, *attensity* was used by a group of early "introspective" psychologists, who believed that they could, working together to study the experience of their minds-in-action, learn to "see" the actual dynamics of human cognition. (Interestingly, their leader, Edward B. Titchener, not only came up with the term "attensity," he also coined the word "empathy" in English; happenstance? Maybe . . . but *maybe not!* After all, true attention and authentic fellow feeling go hand in hand.) These folks did some beautiful and quixotic work—but their program eventually lost out to more mechanistic experimentalists, who decided thinking about thinking was too unreliable to be actual "science" (the new guard stuck to treating brains like input-output systems). "Attensity" was abandoned—a terminological orphan from a short-lived world in which it seemed possible to explore attention *itself,* rather than just its track-and-trigger operations or effects.

We'd like to go back and get it, and raise it on the proud standards of a new movement of utopian revolution. *STOP DESTROYING OUR ATTENTION!* Because *WE WANT SOMETHING BETTER FOR OURSELVES AND OUR CHILDREN!*

ATTENSITY *NOW!*

There are deep ways, too, that our work parallels the forms of transformative restoration that have actually helped our natural spaces recover from misuse. For instance, aspects of our movement can be likened to the work of bio-prospectors, seed collectors, and restoration ecologists—folks who have been on the lookout for rare, unusual, or marginalized species, and who seek to recultivate natural environments that have been erased. After all, what has happened to human attention over the last century is exactly what has happened to the prairielands of America: a monoculture.

A botanist-Martian, dropped off in the middle of the United States, would likely look around and say something like "Wow, what a lot of corn! Not a very diverse ecosystem down here." And that Martian would be right. Nearly half a billion acres of this country are pure monoculture, an area representing something like the whole state of Arizona . . . *six times over.* Those regions represent unrelenting miles of just one thing, stretching to the horizon—soybean, wheat, or, mostly, corn. But there is nothing "natural" about all this. Once upon a time, those monocultured farmlands teemed with more than a hundred different wild grasses, forbs, and sedges—tallgrass and shortgrass prairies so vast that early European explorers thought they had come upon an actual *ocean* of meadow, full of species never before encountered.

Every time we think about our attention we should remember those prairies—and the monocultures that have replaced them. Because the attentional world we navigate today is *monocultural*: What we talk about when we talk about "attention" is, in fact, just one very narrow and specific (if also, now, totally dominant) species. It is the kind of attention that has been cultivated by a century of industrial "attention farming," and we hardly remember all the other attentional modalities that it has displaced. Which attention has carpeted our landscape? The mechanomorphic, instrumentalized, stimulus-and-response kind of attention that became the central preoccupation of the military-industrial complex across the twentieth century. The kind of attention that focused on screen-vigilance (key for radar monitors and the monotonous labor of regulating machines), and the *tap-click* incorporation of human beings into cybernetic systems (key for shooting down enemy craft, or making timed decisions in complex models). This was the kind of attention studied in laboratories, quantified by advertisers, and now optimized by AI-driven search engine algorithms. And it is the kind of attention too many of us find ourselves trying to defend, or improve, or manage in our daily lives. It is the kind of attention that our Screen Time app monitors, it is the kind of attention we fret about when we talk about declining "attention spans" or worry that our kids have ADHD.

But Attention Activism involves shaking ourselves out of the sense that our monocultural attentionscape "is" attention. Some recent commentators have even gone so far as to call for a "rewilding" of attention, and this is a perfect way to think about it. The sweet work involves throwing on some boots and taking a long walk across the carpetfields of clonal *soybean attention* to find a little wedge of old-growth prairieland here or there at the margins. Aha! Look what is still flourishing in the foothills! Beautiful! Here are some *all-too-rare species of attention,* thriving in little ecosystems not yet plowed under by the frackers . . .

For instance, how about the one hundred and seventy-nine dues-paying members of the Phoenix Bonsai Society, who have been quietly working away at their small trees, and recently celebrated their sixtieth anniversary. That's a form of attention that has little to do with search engine optimization. On the one hand, the shallow roots of a bonsai demand daily water, so caring for a trophy specimen requires about as much custodial care as a pet cat. On the other hand, the trees grow so slowly that they achieve their beautiful sculptural form across *decades,* and that means only a few shaping snips every year or so. What an unusual attentional nexus: nearly continuous nurturing, plus the long-long view of envisioning a form that will not emerge until well after your kid has graduated from college!

Or what about that very special mode of attention that involves simply staring out the window? The one that the teacher called "distraction"? *There's* a rare species, and one that begs for a little protection.

Maybe this is the moment to say a few words about distraction. On the face of things, distraction is precisely the opposite of "attention"—at least that's the way the standard denunciations set things up. Attention good; distraction bad. Indeed, this Manichean pairing can be traced all the way back through the key moments in a long history of fretful concerns around personhood, social order, and right living. Patriarchal denunciations of flighty, fashion-addled society women in the late nineteenth century never failed to decry the wandering eye, wandering mind, and general inconstancy of "*la distraite.*" Even more grave, distraction featured as a soul-snatching ruse of the devil across nearly a millennium of Christian devotion. Monks, priests, and the faithful laity, all of whom shared a sense that righteous living required a continuous focus on God, together saw distractibility as a perpetual peril to properly prayerful contemplation. So when a middle school teacher gets on the case of a "distracted" student, a long legacy of disciplining anxiety lies behind such an exchange.

And yet, the work of actually *distinguishing* attention from distraction proves philosophically tricky. After all, when I accuse you of being "distracted," what am I really saying, if not "You seem to be paying attention to something other than what I'd like you to focus on"? Which is to say, "distraction" ends up looking a lot like *unauthorized* attention—or, to put it another way, attention that is not in line with the expectations of whoever's in charge.

Is that all that can be said? And if so, is there actually no difference between attention and distraction? Is the distinction purely a matter of sociology? Of relations of power?

Not at all. So let's sit with the conundrum a little while, for the light it shines on the nature of the authentic freedom at the heart of ATTENSITY.

Attention Activists know the deeper truth that has been plowed under by the industrial combines of our attentional monoculture: attention *isn't* that narrow, determinate, track-and-trigger "focus" that facilitates frictionless integration of humans and machines (and makes bank for the frackers). It literally *is* a lot more than that—it is *many, many* forms of cognitive engagement and sensory awareness; it is faith, love, and life! It is a vast and diverse *jungle,* not a single, highly engineered cash crop. The work of emancipating human attention from the harness of the frackers involves, among other things, again and again INSISTING on that diversity and complexity—on the range and richness of attentional forms. Otherwise we risk trying to "save" or "protect" what is, in effect, the *problem.* What could be more tragic and stupid than creating a "society for the conservation of GMO soybeans"? They're doing fine! If Attention Activists slip unselfconsciously into working to "reclaim" the thin and always-already-instrumentalized form of attention centrally at stake in the attention economy, we will have missed the whole point. So stop worrying about your "attention span," and start thinking about how to put your hand on the brow of that tired old pig, and reteaching it the *long, perfect loveliness of sow.*

To do that, you may need to *meander* a little. You may need to let yourself get *distracted* by what else the world contains (other than the vibrating notifications in your pocket).

It's in this context that we'd do well to sift out the attentional valences of "distraction" itself. Attention Activists need every ally in our asymmetrical battle with the frackers, and double agents are especially useful! Indeed, one quite brilliant commentator on the history of attention, Paul North, has gone so far as to argue that the *deepest and purest forms of attention* actually hide behind the disguise of so-called distraction: It is precisely those moments in which we are most completely "checked out" of the expected attentional patterns and objects—when we are so immersed in a daydream or reverie that, when called to the matter at hand, we truly cannot even recall where we were or how we got there—that the most extraordinary experiences of human attention are achieved.

Realizing that there is something profound, and actively *inspiring,* in North's paradoxical provocation tips us dizzyingly, yet again, to the edge of what language can do. Our terminology can obscure as much as it reveals—sometimes more. And our language for talking about our attentional lives is not especially well-developed. Not only that, it has been systematically distorted by a century of narrowing attention-discourse. The work of Attention Activism includes the delicate and elusive work of staying absolutely true to the complexity, beauty, and specificity of our actual experiences—and declining the flattening simplifications of inadequate vocabulary.

One of the very greatest philosophers of attention, the American psychologist William James, specialized in just that kind of fearless and expansive thinking. His work involved a live-minded preoccupation with the actual "phenomenology" of attention (the study of the experiences of perception, cognition, and "being" itself). And in pursuing that project, he came up with a lovely way of theorizing the "deep distraction" part of an attentive mind. In his telling, the pinpoint focus that locks statically on an object cannot really be called "attention" at all! For him, it's really a kind of inert stultification. Imagine, for instance, the dead-eyed, glassy, vacant gaze of a human being going fully vegetative while scrolling on the phone. Is *that* "attention"?

James himself was dead by the time TV rolled around, but he knew that empty gaze, and knew that regardless how long it lasted or how fully trance-like it became, it wasn't the thing he was trying to describe when he thought about the operation of his own attention at its vitalizing best. On the contrary, James came to believe that in the healthy state a human mind was basically *incapable* of simply "staying with" a discrete object for more than a moment. It was, in his view, in the essential nature of human cognition to be *DYNAMIC,* to be moving, to be constantly in what he called the "stream" of consciousness. The alternatives weren't "focus" or "attention"—they were really just a kind of spirit-death.

Hold up a dime, ask me to give it my full attention. Instantly, I'm wondering what's going to happen next. Sure, I can put my eyes on the dime—and I can even try to "fill my whole mind" with that dime. But in practice, James felt he could simply see, in the actual operations of his own consciousness under those circumstances (like the coiners of "attensity," he called this process *introspection*), that the moment he confronted such a task, his thoughts were immediately and always-already departing from that object: I'm wondering why you asked me to look at the dime; I immediately suspect you're trying to play a trick on me, so I am trying to figure out what you are doing that the dime is supposed to stop me from noticing; oh, and you really should clean your fingernails.

That right there, practically speaking, is what "paying attention to the dime" *actually* amounts to. And James called this out. Having decided that this was an essential feature of normal human sensory cognition (that, to put it bluntly, healthy humans are *incapable* of "attention" in the flat-footed sense of a dead-eyed and unwavering and complete "focus"), James proposed a truly *brilliant* account of what we actually mean when we talk about the beautiful thing that attention can be: For him, the genius of real attention is the ability, again and again, to *bend the path of those moving thoughts back around to the object.* In other words, genuine attention, the phenomenology of human attention, is a kind of *centripetal distraction.* Imagine the lovely looping rings of a spirograph, as it petals out and away and then again back to whatever is at the center of its attentional whorl. (Looks like a flower, doesn't it? *The bud stands for all things . . .*)

It's a compelling vision. It's the vision we claim for ATTENSITY. (And it's the cover image of this book, as well!)

For the Attention Activist, attention will always be the *question,* so we won't allow ourselves to rest easy with any single answer to what attention "is." But in this very restlessness we tip our hat to our forefighter, William James, who knew that giving our attention to attention is indeed a RESTLESS CIRCLING.

And when we circle together? What is that?

THAT is the beauty of a dance, the beauty of a *reel*—that spiraling whirl of merry-go-round delight, the pure play of dizzy joy. This is our movement. So come dance with us, in the *restless circling of TRUE ATTENTION*—the attention that is *free,* and *yours,* and *not yours alone.* The true freedom of an attention that is OURS!

In solidarity,

The Friends of Attention

To learn more, access resources, and join our coalition of Attention Activists, visit: schoolofattention.org

A Valediction

(Forbidding Mourning)

RIP Matthew.
We are coming for the frackers in your honor!

Acknowledgments

Naming anyone is the start of the process of forgetting . . . too many people. So it is hard to start, and hard to stop. The pages above reflect six full years of thinking and talking and doing and drafting by dozens and dozens of individual persons, who have added, pushed, resisted, stayed close, moved off, spun out, doubled down, showed up, and otherwise manifested their particular perspectives and sensibilities in relation to an emerging program of broadly shared critical diagnoses, institutional forms, and conceptual visions. Whew! Not everyone is in this list. And there are some names below of folks who did no more than read a draft, offer support at a key moment, share some ideas with us, or help steer the making of an actual publishable thing out of the messy business of movement politics. Feel free to ask them how they relate to the work we hope you just read. Here they are: Anthony Acciavatti, Casey Affleck, Alex Balgiu, Akua Banful, Daphne Barile, Tina Bennett, Larry Berger (both of you!), Kyle Berlin, Donica Bettanin, David Burnett, Francesca Burnett, Leo Burnett, Raiane Cantisano, Sonali Chakravarti, Julian Chehirian, Christine Chi, Yves Citton, Claudia Claremi, Amanda Cook, Jeff Dolven, Ana Cristina Dos Santos, Joanna Fiduccia, Hal Foster, Brad Fox, Melissa Galvez, Claire Gaudiani (RIP), Christie George, Jahony Germosen, Justin Ginsberg, Jael Goldfine, Jill Goldman, Cheryl Grady, Catherine Hansen, Nathan Heller, Stefanie Hessler, Aaron Hirsh, Brooke Holmes (at Pitt!), Adam Jasper, Stevie Knauss, Mihir Kshirsagar, Will Lamson, David Landes, Kristen Lawler, Irv Loh, Trudi Loh, Tracee Mallamo, Quinn Marchman, Açú Marques, Alice McCrum, Helen Miller, Eve Mitchell, Chris Mole, Carlos Montemayor, Jac Mullen, Len Nalencz, Carla Nappi, Vitória Oliveira, Gwen Olton, Gabriel Pérez-Barreiro, Dominic Pettman, Jesse Prinz, Stefan Pryce, Sal Randolph, Jared J. Rankin, David Richardson, Rebecca Rickman, Anna Riley, Marcus Ryan, Iciar Sagarminaga, Henry Schmidt, Lisa Siegmann, Caleb Smith, Justin Smith-Ruiu, Federica Soletta, Matthew Spellberg, Hermione Spriggs, Sarah Stillman, Stephen Strother, Lane Stroud, Etienne Turpin, Cody Upton, Catherine Willett, Berta Willisch, and . . .

About the Authors

THE FRIENDS OF ATTENTION are a nonprofit coalition of creative collaborators, colleagues, and actual friends who share an interest in "ATTENTION ACTIVISM." The Friends emerged in the wake of the 2018 São Paulo Biennial, where many of the initial community were involved in a program on "Practices of Attention." Responding to an expanding sense of crisis, eighteen artists, scholars, and activists gathered in the summer of 2019 for a weeklong residency, "The Politics of Attention: Art, Time, Technology, Action," where the group began to take shape in collective reading, writing, and interventions. Public projects include: *Twelve Theses on Attention* (a book, film, and print exhibition); critical writing in *October* and elsewhere; and, since 2023, the Strother School of Radical Attention in Brooklyn, New York. There is no "membership" in the Friends. There are friends.

FRIENDSOFATTENTION.NET

D. GRAHAM BURNETT is the Henry Charles Lea Professor of the History of Science at Princeton University. **ALYSSA LOH,** a filmmaker, co-directed the short film *Twelve Theses on Attention*. **PETER SCHMIDT** is the program director of the Strother School of Radical Attention.